Force

Force

What It Means to
Push and Pull,
Slip and Grip,
Start and Stop

Henry Petroski

Yale UNIVERSITY PRESS
New Haven & London

Yale University Press books may be purchased in quantity for educational, business, or promotional use. For information, please e-mail sales.press@yale.edu (U.S. office) or sales@yaleup.co.uk (U.K. office).

Set in Galliard Old Style with Gotham Medium type by Integrated Publishing Solutions. Printed in the United States of America.

Library of Congress Control Number: 2021950015
ISBN 978-0-300-26079-3 (hardcover : alk. paper)

A catalogue record for this book is available from the British Library.

This paper meets the requirements of ANSI/NISO Z39.48-1992 (Permanence of Paper).

10 9 8 7 6 5 4 3 2 1

To Theodore James

No force however great
can stretch a cord however fine
into an horizontal line
that is accurately straight

—William Whewell
An Elementary Treatise
on Mechanics, 1819

Contents

Preface

This is a book about force, the everyday physical interaction between ordinary things—including people—that enables them to stay in place or causes them to move. It is the force of daily human experience, whereby we feel the sensations of push and pull, weight and buoyancy, resistance and assistance, achievement and defeat. It is the force we fight in climbing a mountain, swimming against the tide, doing heavy lifting, trying to open a bag of peanuts. It is the force that enables us to walk and airplanes to fly. It is the force we feel through a car's bouncing, swerving, swaying, and cornering. It is the physical force by means of which we do everything from raising a cup of coffee to opening a jar of jam. Without forces and their effects, we would lose contact with our world.

Thinking about how we experience force in the context of our everyday activities, we can begin to understand its nature and appreciate how it has been harnessed for our use and pleasure. Force is also a physical link between human beings. The feel of a handshake—whether a firm grip or a dead fish—tells us about the personality and mood of strangers and old friends alike. An open hand may give us an encouraging pat on the back; a clenched fist, a punch in the gut. Through the tender touch of a finger, the soft caress of an arm, or the peck of a kiss, we can convey to another a sense of safety, comfort, and love. Although forces of contact may be mechanical phenomena, in human hands they can be the stuff of war and peace.

Forces themselves may be invisible to the eye, but they are sensible to the touch. Although we experience this countless times as we go through our daily routine, it has become so familiar that we seldom stop to reflect upon it, which is why I have written this book. In it, I have focused not so much on physics as on the physical. I have tried to capture in words what I feel with my senses. I believe everyone already does feel the forces I describe herein; I want to encourage readers to feel them more intensely and appreciate them more broadly in the context of living. There are many ordinary activities in which we engage that involve background forces of which we may not even have been aware. I bring these to the forefront to reveal what we have been experiencing without knowing. I want to heighten sensitivity to force as a means of better understanding our interaction with the world of things. I want readers to feel the forces when they use a pencil to scratch out a note on a scrap of brown paper and feel how they can do it effortlessly with some pencils and not others. I want the reader to recall with me the feel of a helium-filled balloon tugging at my wrist and relive the sense of joy all children discover in a sandbox and a playground. I want us to recognize that as adolescents we discovered new sensations of force in afterschool activities, as adults in adventure, and as seniors in doing things we never thought we would. This book also asks readers to imagine themselves in the place of a caryatid bearing the weight of a Greek temple's architrave and pediment and to play the role of a critical component of a modern bridge spanning a river.

I am an engineer, and I see forces everywhere and feel them in everything I touch. Maybe this results from my favorite courses in college, those that taught me how to recognize the forces that act between parts of a structure or machine, making it remain intact or causing it to move. Learning how to isolate forces conceptually, analyze them rationally, and feel them viscerally opened up to me a new way of thinking about how different parts of experience fit together. Physical force began to look like a linchpin to understanding much around me. In graduate school I began to understand more deeply how force manifests itself in both construction and destruction: the same forces that engineers tame to build can also be harnessed to destroy.

As I began to see forces at work in and around my own body, I saw the interactions of people in society as a manifestation of some kinds of forces between and among them. These forces were less physical than psychological, but their effects could be just as effective in the movements they engendered. For me, the concept of force provided a metaphor for understanding both interpersonal and societal behavior. Whereas applying a tangible force to an inanimate object could enable a person to move it or stop it, the intangible forces of charisma, persuasion, and influence could enable a demigod to initiate and sustain positive social, political, and cultural movements among peoples—or a devil to reverse them. This book provides a guide to understanding how the physical world works in the context of normal everyday activities as well as in unusual circumstances when unfamiliar forces may produce counterintuitive results.

Prologue

Things We Feel

For as far back as anyone knows, humans have interacted with each other and the world by means of the five senses: hearing, seeing, smelling, tasting, and touching. Surviving in a primitive world must have meant having all five on heightened alert at all times. Our prehistoric ancestors had to hear a predator moving through the underbrush toward them, see an attacker before it reached them, smell the air for an approaching grass fire, taste the danger in rotten food, and feel the presence of a poisonous insect crawling down their back. Losing any one of the defensive senses would put one at an elevated risk for destruction.

Socialization brought different threats, sometimes in the form of human-on-human violence prompted by the instincts of selfishness, dominance, and survival itself. Antisocial acts such as deception, cannibalism, and war might even arise. Evolving senses serving a receptive brain became a necessity for an individual to detect the tell-tale signs of a wolf in sheep's clothing. Sounds of people quietly plotting, the sight of them acting suspiciously, the smell of something in the air, the taste of something in the water, the feel of something funny going on raised a red flag. Words and phrases we now call idioms and clichés have their origins in the straightforward experience of the senses.

The industrialization of civilized society may have been revolutionary, but we owe our comprehension of the world and its proper functioning to our evolutionary perception of it. As tools, devices, machines, and other made things and systems proliferated, humans had to recalibrate their senses.

It became necessary for users of technology to know when its implements were about to present a danger: hearing the knock of an improperly fueled engine, seeing the wear on a weakening machine part, smelling the scent of overheated gaskets, tasting the smoke of burning oil, and feeling the vibration of an unbalanced wheel. Understanding the significance of such signs came from experiencing the effects of the forces involved and distinguishing the good from the bad.

In the development of modern medicine, the goal of enhancing our five senses played as important a role as the introduction of surgical instruments and the development of medicines. The stethoscope amplified the sounds issuing from the heart, lungs, and digestive tract; X-rays enabled a physician to see beneath the skin without having to cut into it; analytical tests of biopsied and autopsied tissue supplemented the detection of rot by the nose; chemical analysis replaced the need for a physician to taste a patient's urine; magnetic resonance and other imaging advances tell a doctor much more than palpating an abdomen can. But these technological developments took time, first to be imagined, then to be prototyped and fine-tuned to the point where they were highly reliable.

The development of any new technology must necessarily occur within the bounds of the laws of nature. Hence it is a given that whatever physical, chemical, electrical, or other forces are involved, their exploitation cannot be expected to exceed scientific possibilities. In a Newtonian universe this means that it is futile to pursue a perpetual motion machine. But research and development takes place in a grander context than one of hard physical force; it is also subject to the softer forces imposed by ethics, morals, judgment—none of which is easily and unambiguously defined by laws and limits.

In early 2020, transmission of the Covid-19 virus by physical contact became a growing concern, and the act of touching became something to be avoided. We were advised to eschew shaking hands, hugging, and other timeless expressions of community, friendship, and even love. To replace these reassuring forms of interaction, we pumped fists, clicked elbows, tapped shoes, and even bumped bums. The collections of individuals known as cliques, clubs, klatches, neighborhoods, communities, cities, counties,

countries, and cultures were expected to reimagine what constituted civilized society.

The five senses continued to be reliable indicators of what was in the
air. They were the first responders, the early detectors of symptoms. It
became evident that an infected individual would have difficulty breathing, which could be heard by the unaided ear; have a runny nose or chills
easily seen by the naked eye; have a diminished if not absent sense of smell
and taste, which discouraged eating and led to noticeable weight loss; and
have a fever, which could be felt by the touch of a hand to the forehead.
Sometimes breathing was so compromised it had to be assisted by a ven-
tilator. Visitor access to hospitals and other medical facilities became increasingly restricted. To get past the front door of a clinic I had to have
my temperature taken with a touchless thermometer, use a dollop of hand
sanitizer from a touchless dispenser, maintain social distancing by standing on marks on the floor, and put on a face mask without touching my
face. This made some of us feel that we were increasingly becoming de-
sensitized to our sense of touch generally.

Social and physical barriers established to guard against transmission of
the coronavirus are epitomized in the face mask. In the Middle Ages, dis-
ease was believed to be transmitted through a miasma, the foul-smelling
air emanating from rotting flesh and other disease-laden sources. To protect themselves from the Black Death, so-called plague doctors wore nose
coverings in the form of a bird's beak filled with fragrant herbs, flowers,
and the like to block the miasma. These doctors were depicted wearing
black cloaks and hats and holding a stick, which enabled them to examine
victims of the plague without touching them directly. The stereotypical
plague doctor was a Dr. Beak, and he became a symbol of death and dis-
ease generally.

It was not until the end of the nineteenth century that both the nose
and the mouth of some medical personnel were covered in medical settings,
but the practice remained far from customary for decades. In the meantime, World War I prompted more conscientious use of medical masks
in the United States and Germany, but the nose often remained exposed.
Research into the composition of masks led to the use of washable and
sterilizable ones in the 1940s. It was not until the 1960s that the United
States initiated the use of masks made of paper and fleece, and use of the

Seventeenth-century plague doctors wore a nose mask filled with aromatic flowers, herbs, and spices to perfume the contaminated air they had to breathe; they carried a stick to maintain their distance from the afflicted and deceased. *From an engraving by Paul Fürst (ca. 1656) via Wikimedia Commons, distributed under a CC BY-SA 3.0 license by user lfwest.*

disposable kind became widespread. Throughout the pandemic of the early 2020s in the United States, the procedural or surgical mask became the default nose-and-mouth covering. It appeared at first glance to provide reasonable protection against contracting and transmitting the coronavirus, but it had some serious limitations relating to its physical construction and function.

The surgical mask takes the form of a multilayered, horizontally pleated rectangle measuring approximately three and a half by six and a half inches. The nature of the manufacturing process fixes these dimensions, so the edges of the rectangle cannot stretch to provide an optimal fit on every size face. However, the pleats do allow the filtering material to expand vertically to cover both nose and mouth and in this configuration puff out to provide breathing room between mask and face. Better masks incorporate an easily bent length of wire or metal strip into the top edge of the rectangle so that it can be molded to conform, but not perfectly, to the contours of the nose. Elastic cords anchored in the vertical edges are looped over the ears to pull the mask against the face. However, because the distance between the anchor points of the cords is fixed, the sides of the mask cannot stretch but they can contract, which often allows a gap to open up between them and the cheeks of the mask wearer. This pair of gaps, along with those around the nose, can reduce the effectiveness of the mask to contain air exhaled by the wearer and allow outside air to enter in ways other than through the filtered rectangle. If such shortcomings are to be corrected by redesigning the common mask, it will almost surely have to be by redirecting the forces in and on its fabric. For example, shorter but more stretchable vertical sides would allow the mask to fit tightly against the cheeks of small-framed persons while being able to stretch to conform also to longer-faced people.

If the standard mask does not fit all adult face sizes equally well, and if improvements involving a redistribution of forces on the mask are not forthcoming, perhaps different-size masks should be made more readily available. During the pandemic, I encountered no hospital, medical center, or doctor's office that offered anything but a one-size-fits-all mask. Although an ill-fitting mask is likely better than none at all, the availability of masks in even a limited range of sizes—say, small, medium, and large— could go a long way in improving their effectiveness. It might also eliminate another obvious deficiency in the single-size standard issue, and that is the fact that it is prone to slip off the wearer's nose, effectively rendering it ineffective. The cause of this slippage can easily be traced to the physical forces at play.

When a mask whose vertical sides will not stretch reaches from over the nose to under the chin, flexing the jaw to talk effectively, chew aggressively, or yawn widely pushes the chin against the bottom of the mask, which in turn pulls it down the bridge of the nose. If the ear loops do not pull it tightly against the bridge of the nose, the mask will have little gripping power there and will progress down the nose until it slips off the tip and exposes the nostrils. Health care workers and others who were annoyed by this failing devised clever ways to pull the mask tighter against their face. One is to twist the elastic cords into a figure eight or knot them behind the ears, thereby effectively increasing the amount they have to stretch and so increasing the force with which they pull. The problem might also be solved by redesigning the mask itself, such as by making the cords shorter to begin with, making the sides of the mask shorter but more stretchable, or adding more pleats to allow the mask to be more expandable vertically to accommodate the greater movement associated with longer faces and wider-opening mouths. Some fashionable alternatives to the surgical mask do fit more properly, but other mechanical problems remain.

People who wear eyeglasses experience a special problem when their mask does not fit tightly around the nose. Warm, moist exhaled air can escape through gaps along the mask's top edge and fog up the lenses. This can render the mask wearer momentarily sightless, an especially dangerous condition for someone walking across a road or working at a delicate task. A mask that overlaps the bottom edge of the glasses will be pulled farther from the face and so exacerbate the problem, which is why it is important to have the glasses overlap the top of the mask. Fogged-up glasses have long been the bane of surgeons in the operating room. Fog-resistant lens coatings help, but they can be easily scratched and become delaminated. Some resourceful surgeons prevent condensation on their eyeglasses by using surgical tape or adhesive bandages to bind the top of the mask to the part of the face beneath the eyes. Understandably, the solution did not become commonplace outside the operating room. People who wear contact lenses can also be adversely affected by the current of air issuing from the gap between their face and the top edge of a mask. This warm, moist environment can promote bacterial growth and eye infections. Disposable soft contact lenses especially can get dried out and lead to irritation that might have to be treated by an ophthalmologist.

Wearing two masks has on occasion been advised, not only because it increases filtration capacity but also because it can press the inner mask more tightly against the face in places where it is prone to gapping. Of course, even this may not improve the effectiveness where the wearer is a man with a bushy beard or a woman with gaunt cheeks.

Professional and amateur inventors are always looking for new challenges, and many a patent is issued for an improvement in even the most common of devices. It is almost a sure bet that marked improvements to the surgical mask will become available before the next pandemic strikes. And these new and improved masks will very likely rely upon innovative ways of controlling and exploiting the physical forces that keep the prophylactic device firmly in place. Many hopeful inventors will no doubt focus on the parts of the standard mask that have been identified as being responsible for the problems—overall shape, inextensible perimeter, pleating, ear loops—to come up with improved designs. Some more adventuresome inventors may identify previously unidentified weaknesses in the design and devise a revolutionary model that will be inexpensive to manufacture and sell. Such are the dreams of inventors.

A critical piece of antivirus technology came to the fore in late 2020, when vaccines became available. In the course of promoting vaccination as a positive act, celebrities and politicians were videotaped receiving their shot without wincing. Many of the video clips shown on the evening news featured a closeup of the needle entering the arm. These videos demonstrated the many different ways a vaccinator could administer the vaccine, and collectively they provided a short course in aspects of force. One method, which seemed to be followed by health care workers not used to giving shots, was to place the tip of the needle on the target spot and slowly push it against and into the skin. This produced a visible bowl-like depression in the arm, showing clearly how the skin at first resisted being pierced and then yielded under the force of the needle and rebounded. On the other extreme were the more experienced nurses who drove the needle quickly and deliberately into the arm. It happened so fast that there was no visible change in the topography of the skin. When the needle is jabbed, it imparts a dynamic force that pierces the skin before there is time for it to distort to any noticeable extent.

What causes a mask to stay on the face and what causes one to slip and gap is mechanical force, the subject of this book. Vaccinating a person is possible because forces enable the vaccine to be pulled out of a vial by a syringe, to be pushed into muscle by a needle that itself is pushed through the skin. In the mechanical world of artifacts generally, a key enabling factor is force, which depends heavily upon contact and touch. Focusing our other senses on the one of touch offers a fresh way of experiencing and understanding the world.

In his essay "Nature," Ralph Waldo Emerson called for humans to have "an original relation with the universe." In laying the foundations for transcendentalism, he argued that studying nature leads to a greater understanding of everything. Engineers may not aspire to develop an entirely new philosophical movement, but they can believe that studying the forces of nature and their effects can lead to a broader view of how the physical world works and, by extension, a greater comprehension of its social and cultural structures, as well as the metaphors that bind together the physical and emotional. Although Henry David Thoreau, another transcendentalist, asserted that he "was not born to be forced," he must not have hesitated to force his hammer to drive the nails used in holding together his cabin at Walden Pond. Reflecting on nails and what makes them work can be as enlightening as studying softer aspects of nature, and it can be just as satisfying.

1
Pushes and Pulls
Sources of Forces

What is force? Is it something we need science to define for us? Or do we simply know it when we feel it? But what does it mean to feel a force? Does the sensation merely confirm the existence of a concept we call by that name? Or does feeling a force evoke an inner sense of our place in the universe and our connection, however remotely, with everything else in it? Force, both in its presence and in its absence, provides a means to a better understanding of our connection with the physical world in which we exist and go about our ordinary and extraordinary activities.

In a physical sense, to touch something is to apply a force to it. Musicians in an orchestra sit still and silent until the score calls for them to play their part. When that happens, strings, reeds, winds, and percussions come alive as musicians push and pull; bow and blow; pick and pound; mute and mash their instruments to induce and arrest vibrations in everything from a single violin string to an expanse of drumheads. The object of the action is to turn force into movement into resonance into sublime sound. After we leave the concert hall, the strains of a symphony still reverberate in our heads. The effects of force are moving and long lasting.

The sound of a human voice also relies on mechanical forces. Air pushed out of the lungs moves across vocal cords and causes them to vibrate and emit sound waves (the way wind blowing across power lines can cause them to hum). As the sound travels outward through the speaker's mouth, it is modified by the shape and the position of the tongue and lips, as well as by the nose (the way a wind instrument has its pitch modified by altering a vibrating column of air).

Sounds and forces go together: the splat of spilt milk; the wham of a hammer; the pop of a balloon. If we listen closely, we can hear even softer sounds of scratching, sliding, and scraping in everything we do: pushing a pencil across a sheet of paper, pulling a shelved book from between two others; running our fingers across the stubble on a cheek. On a quiet night, we may even hear the sound of our breathing and the pounding of our heart.

The forces that produce the vocal sounds enable us to pronounce the vowels and consonants we combine to articulate syllables, words, and sentences that constitute communication, whether to order a hamburger and fries or to cheer on the home team. Among the most effective means of verbal communication is the well-structured lecture. Paradigmatic examples from ancient times include Plato's *Dialogues,* which preserved the thinking of Socrates and his method of eliciting thoughtful responses from his listeners by asking evocative questions. In the Renaissance, Galileo employed the Socratic method in his *Dialogues Concerning the Two Chief World Systems,* in which he treated the motions of the planets around the sun, and also in his *Dialogues Concerning Two New Sciences,* in which he considered how materials resist breaking under force and how objects move toward the Earth.

The lecture format in which a teacher stands in front of a class and reads from an authoritative text as the students take notes on its content became a staple of university education during the Middle Ages. It persists to this day in seminars and congresses where graduate students and professors, especially those in the humanities and social sciences, literally read their papers aloud. In the sciences and engineering, by contrast, the lecture has traditionally been delivered extemporaneously and simultaneously with the lecturer drawing diagrams, deriving equations, and plotting graphs on the blackboard. Historically, images relevant to the topic of the presentation have also been projected via lantern slides, the forerunner of today's PowerPoint presentation. In the nineteenth century, especially, the popular lecture as explication and explanation of scientific principles was accompanied by real-time demonstrations and experiments conducted on the table behind which the lecturer stood.

In Victorian England lectures became very popular outside an academic setting, and they were delivered to eager audiences of people from all walks of life. In 1868, Thomas Huxley, the biologist, anthropologist, and staunch supporter of Darwin's theory of evolution, delivered a lecture to a working-class audience in Norwich. He began his famous "On a Piece of Chalk" by holding up a lump of carpenter's chalk and proceeded to discourse on the geological history of Britain, including the nature of the White Cliffs of Dover and the Chalk Marl formation beneath the Channel between England and France. A century later the experimental psychologist Dael Wolfle described Huxley's lecture as an outstanding exemplar of "the art of explaining in compelling and understandable terms what science is about." To Huxley, it was "nothing but trained and organized common sense," and so a lecture on science should be accessible to a general audience. The same may be said of engineering.

Thomas Huxley was not alone as a popularizer of science. Michael Faraday, the British scientist who made fundamental contributions to the field of electromagnetism, is also famous for the series of popular lectures he established in 1825 at the Royal Institution of Great Britain. Except for a hiatus during World War II, these Christmas Lectures continue to this day. Faraday himself delivered them during nineteen seasons, including every one over the decade 1851–60, on the topics of chemistry, electricity, and force. Londoners of all ages and genders were enthusiastic attendees. In 1848, he gave his famous course of six lectures on the chemical history of a candle, in which he discussed the nature of the flame as a candle burned and flickered between him and the audience. He spoke in detail about a candle's wax, wick, smoke, and the differently colored regions of the flame. Over the 1859–60 holiday period, Faraday gave "a course of six lectures on various forces of matter and their relations to each other." In his opening remarks to the first lecture, which dealt with the force of gravitation, he addressed specifically the younger people in the audience and, "as an elderly, infirm man," told them that preparing the lecture enabled him to "return to second childhood, and become, as it were, young again among the young."

But Faraday also remained firmly in control of his adult faculties as a scientist who made major contributions to chemistry, electricity, and magnetism. In each of these fields he relied heavily upon experiments, which

Michael Faraday employed a variety of props and artifacts when he delivered his Christmas Lectures at the Royal Institution of Great Britain. The lecture pictured here, from a sketch made by the Scottish portrait painter Alexander Blaikley, was delivered during the 1855–56 season. It was attended by H.R.H. Prince Albert and his sons Edward, aged 14, and Alfred, aged 11, all sitting front and center. © *Royal Institution / Bridgeman Images.*

necessarily required pieces of laboratory apparatus. If he did not have a particular item, he could invent it. When he required a supply of hydrogen at hand for an experiment in 1824, he fabricated a bladder out of the natural rubber called caoutchouc, an "exceedingly elastic" material known today as latex. The ad hoc device earned Faraday credit for having invented the toy balloon. When he lectured to the public, he stood within the recess of a table on which were spread an assortment of props and pieces of apparatus, none of which looked terribly complicated or intimidating, but neither was their purpose necessarily self-evident. In fact, they were integral to the lecture, and their function and relevance to the topic were revealed as the lecture progressed. He used these physical artifacts to introduce, demonstrate, and support the ideas he was expressing in words.

His showmanship toward a serious end pulled in an eager audience. His actions drew rave reactions.

Throughout his lectures on the forces of matter, Faraday sometimes used the terms *force* and *power* interchangeably. To him, heat and water had power, in that they had the ability to effect change, but force was the more general term. He insisted there were not a confusing number of forces to master, noting that it was "wonderful to think how few are the powers by which all the phenomena of nature are governed." And he needed relatively few props to make his point. Among those he used to explain gravitation were a pendulum, a balance, a beaker, and a toy he "happened to see the other day" to make the concept concrete for his audience.

To illustrate the concept of center of gravity, Faraday dropped a handful of shot to show how the pellets were individually subject to the gravitational pull of the Earth and so ended up scattered about the table. After collecting them into a bottle to form a congregate mass, he explained that there was a single point at which all their gravitating power was concentrated. Then he made the concept more personal. He asked the audience, "What do I do when I try to stand on one leg?" He answered the question using his own body as an exhibit. He stood on his left leg and pointed out that he was leaning his body to the left and bending his right leg into a position to keep his body's center of gravity in line with the vertical pull of gravity passing through his foot—his point of support. Faraday knew that describing this in words alone would not be easily understood; by accompanying his words with a live demonstration he made the concept something to which his audience could relate and try for themselves later in the evening.

Returning to standing on his two feet, Faraday turned the audience's attention to the toy doll he had recently acquired. Its upper body was in the form of a person, while its lower, hemispherical one looked like that of Humpty Dumpty. He showed how the toy would not stay put when laid down on its side—something it should be expected to do "if it were uniform" in composition. Faraday explained that he was "perfectly sure, without looking inside," that the wooden figure contained some kind of leaden insert near its bottom that made its center of gravity be near the lowest point of the ovoid base on which it stood. By nature of its geom-

In a lecture on gravitation, Faraday used a weighted toy doll to demonstrate that whenever it was laid down on its side (*left*), its low center of gravity caused it to return to an upright position (*right*). *From Faraday*, The Forces of Matter.

etry, whenever the doll was pushed sideways, its center of gravity was raised. As soon as the pushing force was removed, the center of gravity was free to fall to a lower position, which corresponded to the doll standing up.

Faraday next demonstrated how difficult it was to balance the doll on a point, such as that of a pin or toothpick. However, by burdening the "poor old lady" with a drooping wire yokelike device, each end of which held a lead shot—something akin to the yoke a milkmaid might carry or the long pole a tightrope walker might employ—the doll could easily be balanced on a point. This is because the center of gravity of the system comprising the doll and her burden was now located below the point of contact and she would tilt to the angle that made it coincide with a point on the shaft that resulted in balance. In this book I hope to follow Faraday's model. At the end of his first lecture, he wrote on the blackboard behind him the heading "FORCES" and below it the "names of the special forces according to the order in which we consider them." The first word on the list was "GRAVITATION," which he wished to consider further in the next lecture.

The twentieth-century Israeli physicist and philosopher of science Max Jammer recognized an "intimate historical connection of the concept of force in classical physics with that of gravitational attraction, although of course the ultimate origin of the concept lies in the muscular sensation associated with push and pull." Indeed, every imaginable action resulting from human muscle power—walking, running, bending, twisting, spinning, tumbling—can be interpreted as a set of pushes and pulls. Consider the everyday act of opening a jar of jam. We learn to do this without looking or thinking, but if we do look down on the process we see one hand wrapped around the glass jar to keep it in place as the other hand grasps the lid and twists it off. The twisting action is the result of our thumb and fingers grasping the edge of the lid and squeezing (pushing) it radially inward on diametrically opposite sides and simultaneously pushing tangentially in a counterclockwise direction. When the lid is free of the jar's screw threads, it is lifted off with an upward pull.

According to the chemist William Crookes, who edited the transcript of Faraday's lectures on force taken down by a "careful and skilful reporter," the words were "printed as they were spoken, verbatim et literatim." Crookes also noted that because the talks were directed to young persons, they are "as free as possible from technicalities," and so required no prerequisite knowledge of the subject. But they do deal with more than trivial concepts.

Crookes opened his preface with a question and a proposition: "Which was first, Matter or Force? If we think on this question we shall find that we are unable to conceive of matter without force, or force without matter." Indeed, they are linked through motion. When a net force is applied to a discrete amount of matter, we see it move; when that matter is us moving with other than a constant speed in a straight line, the accompanying force is palpable. This is essentially what Isaac Newton declared in his second law of motion, which relates force and mass through the change in motion known as acceleration, a fact of nature expressed concisely and universally in the simple formula $F = ma$.

Faraday neither used nor needed mathematical symbols to convey concepts of force to a general audience, and neither will I. However, it seems worth noting that just as Einstein's iconic equation $E = mc^2$, linking energy and mass through the speed of light, has become as familiar as Alexander Pope's poetic statement of equivalence, "To err is human," so should Newton's $F = ma$ be seen as a distillation of a great idea. Equations and formulas are simply alternate forms of communication. We can think of $F = ma$ as just another way of writing the sentence, "Force is mass multiplied by acceleration." A definition of engineering as the joining of forces and numbers might be expressed as $E = Fn$. Equations are not at all necessary to understand force and motion, but I could not pass up this opportunity to show how effective they can be. For those to whom equations appear no more meaningful than hieroglyphics, let me say that they are little more than ways of conveying universal concepts using as few characters as possible.

With apologies to John Keats, his equating truth and beauty in his "Ode to a Grecian Urn" might be summarized by an irreverent engineer as simply $T = B$. Admittedly, a profound idea condensed into $T = B$ does not carry the full emotional weight of the classic poem whose words and meter give depth to the thought. The same holds true for $F = ma;$ the fullness of its meaning becomes evident only when it is elaborated upon through examples. Equations and formulas can be looked upon simply as encapsulations of a concept rather than as challenges to calculate a number or solve a problem, in much the same way that Keats summed up the implications of all the imagery, simile, symbolism, metaphor, meter, rhyme, and other literary devices that infuse the forty lines of his ode in its final two: "Beauty is truth, truth beauty,—that is all / Ye know on earth, and all ye need to know."

I will also follow Faraday in revisiting my childhood, for it was then that I first experienced force, felt its effects, and established a relationship with it. I expect that my juvenile reveries were not unique. I am told that when I was an infant, my mother often wheeled me in a baby carriage to Brooklyn's Prospect Park, where she met with other new mothers to gossip, chat, and compare notes. En route, the only view I would have had from my mobile bassinet was upward, and though I would not have been able to see the sidewalks and streets along which we traveled, I must have felt the effects of being transported over their cracks and down from and up onto their curbs. As I have observed babies doing, I must have enjoyed being rocked and rolled over the cracked concrete. I might have seen my mother pulling up and pushing down on the handle of the carriage when it had to be inclined to negotiate curbs, but it was only when I was old enough to walk beside the carriage in which my brother was being pushed that I could see how the motions correlated with the forces. As I grew older, I ranged farther from my mother and began to gain new perspectives on the world. In time I looked down from the upper branches of trees and compared notes with my fellow climbers.

When I had to play alone, I could do so with an imaginary friend whom I could think of as a buddy pushing me on a swing, pulling me downhill in a wagon, or causing me to trip and fall, actions that a parent would recognize and I would eventually know as the elemental forces of push, pull, and gravity. Children also love nursery rhymes, and after hearing one so many times we come to recite it by heart. As we grow up, we may wonder exactly what the sing-song rhyming words might mean. Was Humpty Dumpty pushed or pulled off the wall? What made Jack fall down and Jill come tumbling after? And what about London Bridge? Why was it always falling down, no matter how it had been built back up? We can often answer our own questions by recalling our own experiences. I certainly could as a child. But I always had more questions than ready answers.

In grammar school, I recall being taught arithmetic but not physics. There were no visual displays of even the simplest of machines the way there were pull-down maps for geography. There were no home projects involving hammers or levers, pulleys or wheels, pushes or pulls. We were not asked to make a poster showing the forces involved in such common activities as walking, jumping, climbing, or pulling a wagon. It was evidently more important to know the capital of South Dakota and what industrial products were made in our home state than how and why they were made and grew. Our minds were pulled toward the acceptance and rote memory of facts rather than pushed to an understanding of how the physical world works, even though it was constantly working with and against us through the medium of force.

With our family growing, my parents moved us from an apartment near the southwest entrance to Prospect Park to a house on Park Slope. As a young child there, I loved to walk with friends up the hill to the park and enter it by climbing over the surrounding stone wall. We seldom went over it directly, for we accepted the implicit challenge to walk along its cap. It took an effort to pull myself up onto the wall, and in some places I needed the help of a friend pushing from below. Once up there, I would walk along it as far as I could for the sheer joy of balancing on this imagined precipice. Sometimes I did not make it from one entry gate to the next because, like Humpty Dumpty, I fell off the wall, mostly because of my inattention. Is that why Humpty fell? The rhyme never does tell us. Perhaps that is the charm of nursery rhymes: they leave so much unsaid and so prompt us as children to develop our imaginations.

Inside the park there seemed to be endless expanses of hills and meadows. Experiencing these made me wonder why Jack and Jill went up the hill to fetch a pail of water in the first place. The park's ponds and lakes seemed always to be at low spots in the terrain. On nearby paths, water would well up in potholes formed where hexagonal paving stones were missing. We could see for ourselves playing in the gutters of Park Slope that water ran downhill. In the park, we saw in the drinking fountains the spout of water arcing like a Spalding thrown to a friend and eddying around the drain the way an errant ball rolled into a storm sewer. Perhaps we should have been taught in school that the source of our city's drinking water comes from rivers and streams located upstate and that upstate did not only mean north of the metropolitan area but also suggested that the land was at a higher elevation. The water came down to New York City through buried aqueducts, only to be raised again by mechanical pumps to reservoirs and rooftop storage tanks from which it was distributed to homes and apartments, where it flowed from faucets with considerable pressure. But what was that force? It must have been a push, for who could imagine water being pulled? That seemed impossible.

There are in fact forces that do not require the kind of physical contact that pushes and pulls do. Perhaps the most familiar of these is the mysterious force of gravity. Even Isaac Newton grappled with the concept of gravitation and how a force was transmitted between objects that were not touching each other. Is the ubiquitous gravitational pull between orbs fundamentally different from the push between two colliding balls? Yes and no, for we know that it is gravity that keeps us from drifting away from the surface of the Earth but cannot keep a balloon on a string from flying away from a child's grasp.

2
Gravitation

The Heavy Force

The contractor in Maine advised us to be anywhere but home on the day his crew was scheduled to install new shingles on our roof. He warned us that the hammering would resonate throughout the house, and it might sound as if we were being pelted by hailstones the size of baseballs. Neither my wife, Catherine, nor I took his advice. Her study is two floors down, and she felt that she could wear noise-canceling earphones. My study is in the attic; the roof is my ceiling, and that day it was to be the drumhead of my space. Given my tendency to see and feel force in all things, I relished the opportunity to experience its effects in a new context.

I had no direct view of the goings-on outside, but from the sounds I could picture every action and the forces involved: First, a couple of trucks drove down our gravel driveway and stopped beside the house. Next, judging by the number of doors that slammed, at least four men got out. Metal-on-metal sliding and ratcheting signaled that two aluminum ladders had been removed from the truck's rack, extended, and set against the eaves. Rubbing sounds spoke of ladders flexing under the weight of a man bearing a burden. Heavy trudging footsteps moving toward the ridge indicated workers carrying fifty-pound bundles of shingles uphill. A loud, flat thud evoked a muscleman slamming another onto the trampoline known as a wrestling mat. Softer and faster footsteps meant that the worker was coming back down the slope to take another bundle up to where it would be needed as the hammering progressed. And so it went.

Throughout, I was thinking of each action in terms of force. Along with the repetitive sounds there was a silent partner magnifying the effect. Gravity itself is a dominant but mute force, responsible for giving weight to ladders, hammers, shingles, and workers alike. It was the force against which their muscles had to fight constantly. It was gravity that threatened to make the workers and unsecured shingles slide back down the incline. But gravity also kept the bundles where they were placed and the men where they stood, because the texture of shingles and thick-soled shoes gave them purchase on the slope.

After the silence of a coffee break, the real work began and the house began to resonate. Had the workers planned on using guns fitted with magazines of nails, there might have been a boringly repetitive rat-tat-tat (pause) rat-tat-tat (pause) sound, the pauses occurring only because the roofer had to reach for another strip of shingles to put in place. Even an aficionado of force would soon have tuned the rhythm out.

Fortunately, these roofers were of the old school. They wore belts with loops for tools and pockets full of loose large-headed nails and used wood-handled hammers to drive them one at a time. To do this, the men had to feel among the tangle of nails in their pocket, pull out a few and manipulate one of them so it was placed point down and perpendicular to the shingle. This being done in a split-second, they spent the rest of a second swinging the hammer to pound the nail home.

Men like the ones on my roof are likely to engage in a popular form of skill and entertainment while on a busman's holiday at the state fair or a fundraising event. In one version it involves driving a nail into a board with as few hammer blows as possible. A variation requires a competitor to drive as many nails as he can in a specified period of time. The nail is considered driven when its head meets the surface of the board. In German-speaking countries and communities, such contests are known as *Hammerschlagen*. They attract crowds because each entrant has a slightly different style, and everyone hopes to see a strongman drive the nail with one decisive hammer blow.

Since the crew on our roof was not a group of robots, each worked with his own distinctive rhythm, and they were never exactly in synchrony. This meant that the hammering was never exactly a rat-tat-tat but was maybe a rat-a-tat-tat or a tat-tat-a-rat or something repetitive but not repetitious. Whatever it was, it held my interest as an engineer. It was the application of mechanical force with a human touch. It was lifting the heavy hammer out of gravity's grasp and letting it down with the levity of playing with that same gravity.

From our earliest days, we are accompanied by Mother Earth's pull, which the poet Alfred Corn has called "the planet's comforting embrace." We may not think of the all-pervasive force as a blanket of reassurance that we are part of the world, part of the universe, but it is. It nestles our body in our parents' arms; it places our head in their hands. With the slightest breeze, gravity helps rock the cradle in the treetop. Whatever we do growing up, gravity is always there helping us practice. It pulls us into place, all the while teaching us, warning us, cautioning us, sobering us to the hard realities of the rise before the fall. It is gravity that challenges us to soar and yet keeps us grounded. This ubiquitous and omnipresent force is both faithful friend and fickle foe.

Although we may not have understood so as children, it was terrestrial gravity that enabled us to jump down from a table or chair and land predictably back on terra firma. For me, it was the strong, silent, and secret playmate who tirelessly pushed my swing up and then pulled it back down in a graceful arc. Indirectly, it enabled me to walk, run, and stop on a dime, because its forceful nature constantly attracted my raised foot back down. Gravity kept me from floating off like a Mylar balloon into the thinner air of dreams, as astronauts have repeatedly demonstrated during spacewalks. Gravity was always around helping me enjoy sleds in winter and slides in summer.

But gravity's dark side lurked around, like a big bully. I ceased thinking of the force as a lighthearted It; I began seeing it as a heavy-handed He. His pull could suddenly turn into a push, causing me to fall off walls and fences, poles and trees, monkey bars and jungle gyms, casting me without mercy against the unyielding ground. Gravity may be a lovable oaf, but like an overly large and fractious family dog, he could also seem relentless in his pursuit. Yet a dog eventually tires of running around; my invisible friend never did. Wherever I went, he would be there waiting. When I climbed a hill or a fence, he would pull my whole body back down, letting me know that I was not alone. When I jumped from the top of a fence down to the ground on the other side, he rode on my shoulders and made the sensation of landing all the more intense. When I rode my bike up and down hills, he was always riding with me as I enjoyed the variations in speed with the changes in elevation and acceleration on rolling and snaking paths. When I pulled my wagon up a hill, he invariably sat in it, making it heavier than I remembered it to be; when I got my turn to sit in it for the ride downhill, he ran ahead pulling it faster and faster. He may have been an invisible friend, but he was never just an imaginary one. Gravity was for real and could always be counted on.

It was Gravity who would play ball with me when no one else was around. I could throw the ball up high and know he would catch it at its apogee, hold it for an immeasurable instant, and throw it back down to me—and I had better be ready because he had a wicked fastball and a deadly sinker. As an older child I learned to throw the ball up at an angle, which was mirrored in his return throw. I could run and catch it if I threw it at just the right angle and speed. When I misjudged or just plain missed, I often blamed Gravity, but I knew in my heart that it was really I who made the error.

What became increasingly clear was that some kind of power—maybe alive; maybe not—was indeed present. When I threw a ball against a wall or a stoop, there was always an audible report from the point of contact signaling that the ball was on its way back to me. This was in contrast to the silence of a ball thrown straight up into the air, which made no discernable mention of its having reached the peak of its trajectory and thus had begun its return journey downward. Something special happens to a ball thrown on a somewhat flat trajectory against a wall: it reports its impact. The sound is that of the dynamic force between a moving object and a stationary surface. It is not unlike the sound of a tossed newspaper landing on a subscriber's front porch—or a deposited coin striking the hollow belly of an empty piggy bank. These are the sounds of interaction between objects. They are forces we can hear but not feel, and they are not exactly within our grasp to fully comprehend or totally under our control.

There are also forces that we feel before we hear any sound associated with them. This was the case with V-2 rockets that rained down on London during World War II at greater than the speed of sound. The novelist Thomas Pynchon called the arc of their trajectory "gravity's rainbow." This is actually a misnomer, because rainbows are circular arcs resulting from

sunlight being refracted and reflected by spherical raindrops, which separate the light into its spectrum of colors. The path of a projectile guided by
the pull of gravity is not circular but parabolic. But poetic license trumps
physical prose.

It was the dark side of the universal force of gravity that seized the imagination of a tenth-generation native of the Gloucester, Massachusetts, area
named Roger Ward Babson. As an engineering student at MIT in the
1890s, he successfully lobbied his dean for a course in "business engineering" and in time developed his unorthodox theories of the stock market,
attributing business cycles to Newton's third law of motion, which states
that for every force there is an opposing force of equal magnitude, or, as
more familiarly phrased, for every action there is an equal and opposite reaction. Babson, who would title his autobiography *Actions and Reactions,*
believed that movement in the market was an effect of gravity—what goes
up must come down. In spite of this odd view of market forces, Babson
became known for his business sagacity when he correctly predicted the
1929 stock market crash and profited greatly from it.

Babson spent his money in some unusual ways. During the Great Depression, he provided assistance to unemployed Finnish stonecutters by
sponsoring them in the task of cutting into boulders located in the vicinity of Gloucester's Dogtown Common inspirational sayings such as "Get
a Job," "Keep Out of Debt," and "Help Mother." These monolithic mot-
toes remain in place today along and near the recreational Babson Boulder Trail. In 1940 Babson ran for U.S. president on the Prohibition Party
ticket, coming in fourth, behind Franklin Roosevelt, Wendell Willkie, and
Norman Thomas.

When Babson's sister drowned in a swimming accident, he blamed the
tragedy on gravity, which he claimed "came up and seized her like a dragon
and brought her to the bottom." When his grandson suffered a similar
fate in 1947, Babson was sent "past the point of reason." He blamed "old
man Gravity," whom he saw not as friend but as foe, for everything from
broken bones to airplane accidents. Babson became committed to doing
something to neutralize gravity, or at least insulate people from its adverse
effects. It was George M. Rideout, president of the very profitable Babson Reports, who advised establishing a foundation. In 1949, the Gravity

Roger Ward Babson (1875–1967), who studied engineering at MIT, became a business theorist and used
Isaac Newton's principle of equal and opposite forces
to predict the stock market crash of 1929. Twenty years
later he established the Gravity Research Foundation to
promote the development of antigravity devices. *Library
of Congress, Prints & Photographs Division, photograph by
Harris & Ewing, LC-H25-31581-B.*

Research Foundation was organized and soon published Babson's mani-
festo, "Gravity—Our Enemy Number One." It began funding research toward developing antigravity technologies and sponsoring an essay contest.

To professional physicists, taming gravity seemed too close to pursuing
perpetual motion and other fringe science, and they did not submit essays.
It was only after Rideout convinced Babson to redirect the essay compe-

tition more broadly toward the subject of gravitation that, even in the face of ridicule of the foundation, the generous first prize of $1,000 was tempting enough to attract young researchers such as Stanley Deser and Richard Arnowitt, who were affiliated with the prestigious Institute for Advanced Study in Princeton, New Jersey, with which Einstein was prominently associated. When the paper on "high-energy nuclear particles and gravitational energy" that the postdocs Deser and Arnowitt had written on a lark unexpectedly won the competition, it brought a flood of media publicity of which J. Robert Oppenheimer, the institute's director, disapproved. However, when the controversy passed, distinguished scientists took to the challenge of writing a winning essay. Among subsequent prize winners were Stephen Hawking, Freeman Dyson, and other prominent physicists, which lent prestige to the annual contest. Babson died in 1967, but the essay competition remains part of his legacy under the direction of George M. Rideout Jr., successor president of the foundation.

While Babson was able to be disabused of his dream of finding a way to insulate people from the negative effects of gravity, he was not from donating antigravity reminders to college campuses. In the early 1960s, in conjunction with a $5,000 grant in support of antigravity research, a five-foot-tall pink granite "gravity monument" was installed beside the Physics Building at Tufts University. The inscription on the tombstone-like monument declared its purpose to be "to remind students of the blessings forthcoming when science determines what gravity is, how it works, and how it may be controlled." The monument became the object of defilement; it was tied down so it would not float away and, on more than one occasion, was allowed to fall by the force of gravity into a grave dug in front of it. It was also resurrected more than once, and Ph.D. graduates of the school's Institute of Cosmology have been persuaded to participate in a mock graduation ceremony that involves kneeling before the Gravity Stone and allowing a professor to drop an apple on their head, presumably to remind them of the reality of gravity articulated by Sir Isaac Newton. The tradition continues to prevail over the antigravity dreams of Babson. In 2001, the monument was put in storage because it clashed aesthetically with the exhibition of a piece of artwork by the sculptor Isamu Noguchi. Among other colleges and universities receiving monuments were Colby, Emory, and Middlebury.

Babson was essentially correct in seeing gravity as a source of free power. As surely as the Welsh poet Dylan Thomas saw life itself as "the force that through the green fuse drives the flower," it is the force of gravity that through the penstock drives the turbines in hydroelectric dams and electric generators in ocean-wave-powered devices. Babson might have been more circumspect in seeking an antigravity machine. Zero gravity is in fact a fiction. The universe is a grand system in which every star, planet, satellite (natural and artificial), asteroid, meteor, comet, and piece of space junk is connected to every other one through the force of gravity. The gravitational pull between a planet and its moons may be stronger than that between the planet and matter at the outermost reaches of space, but however weakly they are attracted to each other, they are connected by the invisible link of gravity.

The International Space Station, which orbits the Earth once every hour and a half, is not free of gravity. If it were not for the Earth's constant pull, the station would fly off on a tangent into space. The misnomer "zero gravity" stems from the fact that although gravity is indeed relentlessly pulling the station and all its appurtenances, apparatus, and occupants back toward our planet, their orbital speed around the Earth and the distance from it—which are not independent of each other—combine to produce a countervailing centrifugal force. That force effectively cancels the gravitational pull, a condition that astronauts experience as weightlessness. Engineers and scientists do not refer to conditions inside the space station as weightlessness or zero gravity; rather, they refer to "microgravity." It is an ideal environment for conducting experiments in which gravity is essentially but not actually absent. Things like eating utensils, which just sit on the table at home, do not do so in the space station; they float about the cabin unless held down by something like Velcro. Astronauts sitting down to a meal must anchor themselves with foot loops or some such things, lest they also go floating off. And remedies for indigestion, such as Alka-Seltzer, that go plop, plop and fizz, fizz down on Earth do not work the same way in microgravity. Water does not stay in a glass, and an unanchored table does not stay put. Tourists in space will have to learn to live in microgravity the way babies on Earth have to learn to live with the full force of gravitational attraction.

3
Magnetism

Telephones and Tricky Dogs

The telephone is an excellent example of how a technology can change from one requiring the direct application of multiple tactile forces to one in which touch is optional. Fin-de-siècle candlestick phones, which remained in use well into the twentieth century, typically required two hands to use, as depicted in countless contemporary films in which a split screen showed a man walking around talking into the base and listening on the separate earpiece and a woman on the other end of the line doing the same. The earliest models had no need for a dial. Just taking the earpiece off the hook alerted a human operator that someone wanted to make a call and prompted the question, "Number, please?" As the population of telephone users grew, this system became impractical; the introduction of the rotary dial enabled callers to make their own connections.

The candlestick-style phone was replaced in the 1920s by a more squat style in which the microphone and speaker were incorporated into a single handset. Since it could be held in one hand, the other was free to take notes and doodle. Like Henry Ford's Model T, it seemed, the desk phone could be had in any color, as long as it was black. In the 1940s, the exterior of the Bell System telephone, which was produced by its manufacturing partner Western Electric, underwent restyling by Henry Dreyfuss, an early practitioner of the newly formed profession of industrial design. But the basic elements of the instrument changed little, and its operation continued to require application of manual forces. A hand still had to grasp and a finger had to dial.

The postwar electronics revolution not only changed the innards of the phone from electromechanical to electronic but, through the use of solid state components, made it possible to replace the rotary dial with a keypad and to introduce compact models like the Princess phone. The forces needed to dial were changed from rotational to translational. The term *touchtone* emphasized the connection between force and sound. Early mobile phones have been described as being the size of a box of Kleenex tissues, but they eventually shrank into flip phones and smart phones with the option of no-touch voice dialing. Browsing the web on a smart phone has similarly evolved from having to press the mechanical keys on a BlackBerry, say, to touching only images of them on an iPhone. Google Assistant, a sort of internet operator, has neither a physical nor a virtual keyboard. There is no need ever to touch it, for it works on voice commands.

Throughout all the evolutionary change in form and use of the telephone and related devices there has been one constant. Magnets are essential to their operations, as they are for electric motors, data storage systems, and other marvels of technology. It is the varying force that sound waves impart to the diaphragm in a telephone's mouthpiece that alters the microphone's magnetic field, which translates into a varying electrical signal that is converted back into sound with the aid of another magnet at the receiving end of the call. While we may not feel the magnetic force directly when using these devices, it is possible to do so in other contexts.

The only telephone I knew as a child was a black Dreyfuss hardwired to the wall. It was a ponderous thing that evoked gravitation more than magnetism. To a child my age, the mysterious and yet unnamed force of attraction and repulsion was embodied in a pair of Tricky Dogs, which I bought for pennies at the corner store. The miniature plastic Scottish terriers came packaged in a matchbox. When it was opened they tumbled out as one, held firmly together like a pair of circus acrobats that had locked arms and legs to roll like a rimless tire around the center ring. Naturally, I tried to pull the pair of dogs apart, feeling through my fingers, wrists, and arms the surprisingly strong force that united them. At first, the bond seemed impossible to overcome, but then, with the sudden sensation of a leash breaking, I felt them part. Holding each one by the scruff of its neck, I toyed with the pair of dogs, alternately moving them closer together and farther apart, feeling the forces of push and pull between them. I learned that I had to hold the dogs firmly, lest one or both of them suddenly leap from my grasp. It was more and more difficult to hold them apart as the distance between them became smaller, and there came a point at which the dogs jumped at each other and became locked together once again. Even as children, we grow accustomed to things falling due to gravity and being moved left, right, and every which way by pushes and pulls of fingers and fists and strings and ropes, but experiencing motion across a horizontal surface with no apparent force between the objects involved was something different. It may have led to contact, but it did not arise out of contact. It felt more like conflict.

By the time I had acquired my set of Tricky Dogs, I had already observed my playmates put other teams through their paces, and I wanted to train mine to do equally well. In fact, each of the one-inch-tall Scotties was mounted on a small bar magnet that was the motive force behind all the action. The magnet under one dog was oriented with its north pole below the head and the south below the tail, with the arrangement reversed in the second dog. Since opposite poles attract and like ones repel, the young trainer could get the pair of dogs to perform their tricks by moving one into proximity to the other and letting the forces flow. I knew that if I put the dogs down on a smooth flat surface—a linoleum floor, a Formica counter, a glass tabletop—I could move one to get the other to respond in a predictable way. With the dogs facing off at a respectable distance from each other, moving one Scottie from side to side caused the other to mimic it by wagging not only its tail but its entire body. When one of the dogs was moved in a circle at a safe distance, the other would turn in place, as if chained to a stake in the ground, keeping its eyes fixed on the tormentor. The little dogs played cautiously with each other, as long as the issue was not pushed. If one of them was sneaking up slowly behind the other and got too close, it would spin around in a flash and meet the other face to face and lock it in mortal combat.

I learned to teach the dogs tricks that were new to me but not to them. Among the standard illusions I performed before audiences of younger children and patient adults was to place one dog atop a piece of paperboard (of the kind that new and newly laundered shirts came folded around) and hold the second dog upside-down against the bottom side of the board.

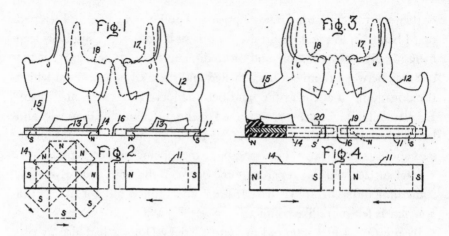

The forces of attraction and repulsion between two small plastic Scottish terriers fitted with bar magnets were suggested in these drawings from inventor Walter Brake's 1941 patent for a "Magnetic Novelty." *From U.S. Patent No. 2,249,454.*

If I could conceal my hand manipulating the bottom dog to direct the top one, it would appear to be moving on its own. The illusion could be enhanced by having people in the audience shout out where they wanted the dog to go and then watch it follow their commands exactly.

A U.S. patent for a "magnetic novelty" was issued in 1941 to Walter J. Brake, an Indiana inventor who worked as a draftsman for General Electric, the corporation to which patent rights were assigned. It was Brake's involvement in the electrical industry, in which magnets have wide application, that made him keenly aware of how they worked and how their interaction could be exploited in a toy that children could buy for pennies.

According to the patent, among the objectives of Brake's invention were to produce "magnetic toys for amusement" and "magnetic novelties of illusory or mystifying character which are moved about without apparent cause." It was not just magnet-driven Scotties that Walter Brake envisioned as embodying both playful and deceptive movements. His patent was illustrated also with a dog and cat pair, who naturally could face off or chase each other. There was also an elephant and a donkey, representing the American political parties. The antics of the magnetized mascots could, according to Brake, simulate the "keen rivalry" between them. Perhaps he envisioned a market beyond children.

Magnetic forces may seem to be supernatural, but to an inventor like Brake they were superpractical. To establish that the most captivating effects of those forces could be incorporated into his toys, he devoted a good portion of his patent narrative to describing his preference for using those kinds of "permanent magnet steels having a relatively high coercive force," meaning materials capable of having a magnetic intensity several times that of ordinary steels. By using such steels, Brake noted, the magnets used could not only be made in short lengths but also would "maintain their magnetism almost indefinitely," making them desirable for incorporating into toys that would hold endless fascination for children, not to mention more serious devices that would not lose their efficacy.

Someone living a half-century after Brake might have chosen magnets made of rare-earth materials such as neodymium or samarium. Rare-earth magnets can be considerably stronger than ferrous ones, some being capable of lifting as much as a thousand times their own weight, but they can also be very brittle and susceptible to corrosion. So the choice of what kind to use in a particular application may not be obvious. Such decisions can be multiplied hundreds of times for an engineer designing something as technologically complex as a satellite that measures the Earth's magnetic field. The engineer who played with even the simplest of toys as a child and experimented with the way they behaved can have a distinct advantage when it comes to imagining how an interstellar space probe will perform as it reaches the outer limits of our galaxy and continues on through the vastness of the universe.

Magnets of all kinds held a fascination for me as a child. I learned that it was an electromagnet (one whose magnetic field is related to the electric current that flows through the wire wrapped around a metal core) that enabled the doorbell to chime, the telephone to ring, and the radio to sing, but mostly I enjoyed playing with magnets that required no electricity to operate. One of my prized possessions was a two-tone (part of it painted red and the other left bare) horseshoe magnet with which I could pick up chains of unlinked paper clips and dangle them like a family of trapeze artists. But mostly I enjoyed just teasing a pair of magnets, like the little Scottie dogs, toward and away from one another, feeling the force between them increasing and decreasing with the amount of separation, but not in a linear way.

I later learned in high school physics that the forces of magnetism, gravitation, and other physical phenomena follow a so-called inverse square law, which means that the force between two objects diminishes in proportion to a fraction whose denominator is the distance multiplied by itself. When I was a child, my curiosity was less quantitative. What materials did a magnet attract? I did enjoy feeling the force of attraction grow as I moved a loosely held magnet closer and closer to a piece of steel to see if I could keep it from being snatched out of my hand before I could pull it away. The magnetic force was a very tactile sensation for me, and I liked feeling it waver as I held two strong repelling magnets very close to each other without letting them touch. The magnets pushed and twisted against each other, an unstable state of affairs. Most of all, I delighted in sensing the very real forces that I could feel but not see.

In fact, we do not see any forces per se. When one Tricky Dog causes another to move, it is not the force between them that we observe but the effect of the invisible magnetic field that surrounds every magnet. In physics we learned to make this field visible by sprinkling iron filings on a sheet of paperboard beneath which we held a magnet. The filings arranged themselves to reveal the field's lines of force along which the iron whiskers aligned like a beard, each behaving like a miniature magnet. If we moved the master magnet quickly enough we could get the fuzz of filings to change direction like a flock of birds in flight. The Earth itself can be thought of as a massive magnet surrounded by its own force field, which reveals itself in causing a compass needle to point to what is known as Magnetic North, which is not the same as True North, the direction toward the geographic North Pole. The compass needle itself is a small magnet. If we were somehow able to distribute millions of little compasses all around the globe, the needles would behave in the same way that iron filings do and produce a visual map of the lines of force.

My youthful experiences with Tricky Dogs, iron filings, and compass needles were of another, simpler time. Today, children grow up in a culture suffused with electronic toys, wireless communication devices, and computer simulations. A smart phone has a compass app that makes it unnecessary to carry what I consider a real one into the wild. We are immersed in a net of magnetic fields, electrical fields, and electromagnetic fields, all of which are invisible but real. Now and then we experience interference among these fields, but generally they and the devices they emanate from sort themselves out to make living in the age that we do full of immediate convenience and ongoing wonder.

My two children were born on either side of 1970, on the cusp of the digital age. Among the most captivating toys of the time was the Fisher-Price School Days Desk, which consisted of plastic letters fitted with little magnets so they could be attached to a metallic easel to spell words. As tends to happen with toys generally, in time parts broke and got lost, compelling the children to make do. Among the things my children did with their diminishing number of plastic letters was to attach some of them to the door of the refrigerator. Since the magnets were just press-fit into the hollow back of each pliable letter, they were prone to being repositioned, if not totally lost. An empty letter did not stick to the door at all, of course, and one whose magnet was pushed too deeply into the hollow stuck only tenuously and, whenever the door was jerked open or slammed closed, was easily pulled to the floor by gravity, which in this case seemed to be victorious over magnetism. For all their flaws, these toys taught our children not only to spell but also to accept the fact that things break and to make the best of it in developing their feel for force.

In 1978, Texas Instruments introduced its Speak & Spell, a breakthrough toy incorporating an integrated circuit and speech synthesizer. The toy pronounced a word and asked the child to spell it by pecking out its letters on a keyboard. The corresponding letters lit up on the display, and the device congratulated the child or showed and spoke the correct spelling. When keys began to break off, my son simply poked a finger into the vacant space to press the formerly hidden switch manually. He seemed to know intuitively that the force he had been applying to a plastic key was transmitted through it to activate something beneath, which could now be worked directly by the touch of his finger. He had developed not only a feel for force but also an appreciation of its power to cause an effect.

Play promotes discovery. The fact that neither gravity nor magnetism requires physical contact to work makes them mysterious, which is what can prompt the question, "Which is stronger, gravity or magnetism?" Although words may fail a parent long removed from child's play, a simple kitchen table demonstration involving a paper clip and a refrigerator magnet will reveal the answer to be, "It depends." The force of gravity will keep the paper clip in place on a table, as long as the magnet is far enough away, suggesting that the pull of magnetism is weaker than the pull of gravity. But when the magnet is slowly lowered toward the clip, there comes a point in the descent when the clip suddenly leaps up and becomes attached to the magnet, showing magnetism to be stronger than gravity, at least in this instance.

On a larger scale, force produced by very strong electromagnets is at the heart of hyperloop technology, the concept invented by the rocket engineer Robert Goddard over a century ago and promoted more recently by the likes of billionaire techno-entrepreneurs Elon Musk and Richard Branson. It promises terrestrial travel at hypersonic speed thanks to the maglev effect, which employs magnets to repel and so elevate the train from the tracks, a phenomenon that no doubt would have excited Roger Babson. The repulsive magnetic force is also exploited to keep the hyperloop vehicle from touching the tunnel walls, thereby eliminating energy-sapping rubbing and scraping contact forces. In late 2020, Branson's Virgin Hyperloop became the first company to demonstrate a working prototype with human passengers on board. The test subjects found the 1,600-foot-long ride that reached a speed in excess of 100 miles per hour to be smoother than traveling in an airplane.

The vast majority of forces we encounter in our daily lives do not act through thin air. Rather, they manifest themselves only when one object actually touches another. In the kitchen table experiment, there is a contact force between the table and the paper clip that keeps it in place. This force is equal to the weight of the clip. There is likewise a force between the magnet and the hand holding it high. This force is equal to the weight of the magnet. At the same time, the paper clip presses down on the table with a force that is its weight; likewise, the weight of the magnet pulls down on the hand holding it. To an engineer, the table, paper clip, and magnet taken as a group may be considered a system within which the paired forces—including those between magnet and paper clip—cancel one another. But neither the weight of the table nor that of the magnet is paired with any force internal to the system, which means that forces external to it must be present to keep everything in place. Those forces are supplied by the floor pushing up on each table leg and by the hand holding the magnet, both of which are considered external to the system.

None of these forces is seen or heard, but the result of their mutually beneficial effects is clearly on display. We do not have to see, hear, feel, or think about such forces as we go about our daily lives, but the same cannot be said of the different forces that come into play when we want to reposition the table by sliding it across the floor.

4
Friction
House Slippers and Finger Prisons

I slept well last night, thanks in part to the force of gravity. My mind may have dreamed that I was floating in space, but my body remained tightly tethered to the Earth. I lie in bed for a while, reflecting on what role forces will play throughout what I expect to be an ordinary day. Seeing nothing unusual ahead, I sit up and throw my legs over the side of the bed to meet the floor with a sensation of solid contact that reminds me that for every active force (in this case, my feet pushing down on the floor) there is an equal and opposite reactive one (the floor pushing up against my feet). Were this not the case, depending upon whether the floor or my feet exerted the larger force, I might either crash through the floor or be propelled toward my bedroom's ceiling. If I return to dreamland, I might imagine myself floating around the room as people do in Marc Chagall paintings, in which gravity does not have any pull.

I feel for the pair of slippers beside my bed. The seemingly straightforward action of sliding a foot into a shoe is in fact fraught with force. If I push my foot in too gently or too aggressively, I may nudge the shoe out of alignment or out of reach. To get my slippers on most efficiently, I must push my foot simultaneously forward and downward on the innersole with just the right amount of force. What that right amount is depends upon many factors, among them being the nature of the materials of which the slipper parts are made; whether my bare foot is sweaty, swollen, or callused; and the nature of the surface on which the slippers rest and on which I will walk in them.

The mechanical force that makes slipping into a slipper both possible and complicated is known as friction. It can be imagined to be due to the fact that, microscopically, surfaces can resemble the topography of anything from the rolling hills of the Palouse to the jagged peaks of the Rockies. When the high points of one surface mesh with the valleys of the other, relative motion between the surfaces is checked, the way it is when the spikes of a golfer are dug into the teeing area. How long this checking can persist depends upon the nature of the contacting surfaces, meaning exactly what kinds of peaks and valleys there are and how they mesh, as well as how forcefully they might be slid across one another.

Friction is an ephemeral and elusive force, because it manifests itself only in the presence of a prerequisite force and a potential condition. The necessary but not sufficient force is that of contact between two objects, such as a foot and an innersole. The condition is a tendency for the objects to move relative to each other, as is the case when I am pushing my feet into my slippers. In order for me to walk, my slipper and the floor must necessarily part ways, and it is then necessary that a friction force manifest itself. Indeed, it is the horizontal friction forces between the sole of my slipper and the floor that make it possible for me to walk at all.

When I stand up, each foot bears about half my weight. If I lift one foot to begin walking, the other one alone must support all of my weight, at least temporarily. As Michael Faraday had to during his lecture, I must adjust my body to compensate for this asymmetrical means of support lest I fall over sideways, which would likely happen if I did not soon plant my lifted foot ahead of me and so shift my weight to it before moving the other one forward. By alternating this action, I can establish a walking gait, with the force that propels me being the friction force acting on the bottom of my slipper because that is my only point of contact with the stationary floor. It is this force that also helps keep one foot in place while the other extends forward. If there were no friction, extending one foot would cause the other to slip backward, a phenomenon known as conservation of momentum, the principle responsible for the recoil of a rifle. As my extended foot comes down on the ground, weight is transferred to it, which makes friction available to keep that foot from slipping. With that foot planted, I can bring the other forward, repeat the action, and proceed apace. Unless the floor is especially slippery, I can move my feet rather

quickly so that the individual forces of contact and friction, being so familiar, are barely distinguishable or even noticeable.

The maximum amount of friction available depends partly upon how
hard I press down on the floor, which in turn depends upon how much I
weigh and how heavily I walk. The friction force also depends on the nature of the surfaces in contact and the force it takes to break the hold of
one on the other. This force is always a fraction of the one pressing the
surfaces together, for that affects how effectively the hills and valleys are
interlocked. The fraction is known to engineers as the coefficient of friction of the mating surfaces. A slipper sole may be made of anything from
leather to rubber, and a floor from hardwood to carpet. Each distinct
combination of materials (leather on wood, rubber on carpet, X on Y) is
capable of producing a different intensity of friction force, and each pair
of surfaces in contact has a characteristic coefficient of friction associated
with it. A coefficient of 0.6, for example, means that a friction force of as
much as 60 percent of the pressure force between a person's slipper and
the floor is possible. When a hundred-pound person is standing still, the
friction force is zero, because no effort is being made to break the inter-
surface bond, but when the person's leg muscles lift one foot to begin
walking, the slipper on the other foot can develop as much as sixty pounds
of propelling force. This force is what does the pushing that propels the
walker.

When we are walking, the nature of contact between our soles and the
floor is not the same as it was when we were standing in place. For one
thing, unless we are stomping very quickly, it takes time for the pressure
between our slipper and the floor to reach its maximum value, which means
it takes time for the friction force to do the same. In the meantime, unless
we have uncommon control over our leg muscles, we cannot keep our
foot from sliding a bit forward on the floor as we take a step. When any
sliding takes place between surfaces, the proportion of friction that is
associated with the pressing force can drop considerably. Technically, the
coefficient of friction changes from a static value to a kinetic one. This
explains why we may feel steady standing still on a wet floor but less so
when we try to walk on it. In fact, we may feel on the verge of, if not actually be, slipping and falling. When water freezes, it presents a very slick
surface indeed, for the coefficient of friction between a shoe and ice is in

the range of 0.2. This is why sand, cinders, and kitty litter are spread across icy sidewalks; they increase the amount of friction available and so provide safer passage.

In some situations we wish there were less friction. Consider a shipping crate weighing one hundred pounds that we wish to slide across a perfectly level warehouse floor. If the coefficient of friction between the bottom of the crate and the floor is, say, 0.5, then as much as fifty pounds of friction force can develop between them. As we begin to push, the friction force will match our effort pound for pound, and the crate will not budge until we push with fifty pounds of force. As we push through the fifty-pound barrier, the box will slip and begin to slide along the floor with a screech. We will feel and hear this quite distinctly, and the force needed to keep the crate moving will be noticeably less than what it took to get it to start moving. The coefficient of friction between sliding surfaces is always less than that between stationary ones. We can also experience the difference between static and kinetic friction if we are climbing up a hill and suddenly find our feet sliding out from under us. This happens when the static friction available to help us advance up the slope has maxed out and the lesser force of kinetic friction is not enough to hold us in place. In Venice, Italy, this happened on the Ponte della Costituzione, a glass-floored arch bridge designed by the architect-engineer Santiago Calatrava. One visitor walking on it expressed shock at how easily his "sandals glided across the glass." So many people slipped and fell on the bridge that its translucent floor panels had to be replaced by ones made of trachyte, the local stone used on the city's streets.

Whenever a liquid such as water or oil is present, it can fill the valleys of a surface and smooth it out somewhat. Our hands and feet naturally get wet when we take a bath or shower, and so they are prone to slip and slide on tile walls and floors. The patterns formed by the small differences in elevation that exist on the fingers and palms of our hands and the toes and soles of our feet produce the unique prints that identify us as us. The skin's topography does smooth out with age, however, and the process of deterioration can be hastened by dermatitis, chemotherapy, and other factors. While someone with a criminal bent might welcome losing his fingerprints in this natural way, most of us do not. I experienced this directly when my fingers felt increasingly slippery when rubbed against each other. I could no longer snap my fingers, an action that involves an audible slip from a static to a dynamic configuration. It also became increasingly difficult to pick up small items by pinching them between my thumb and pointer. When I tried to turn the page of a book with the flick of a finger, it just slipped across the smooth paper.

The irregular surface of normal fingertips does become more pronounced after being soaked in a bathtub or swimming pool, but eventually it recovers its lower profile. Ironically, there are situations in which a wet finger is preferable to a dry one. This is why some people develop the habit of licking a finger before using it to swipe across a corner of a book's page to turn it. The wet finger on the dry paper allows for the development of a greater friction force than a smooth-surfaced dry finger does. For the same reason, a bank teller who manually counts a lot of paper money wets a fingertip on a moistening pad or wears a rubber device that increases the coefficient of friction between the finger and the paper and so produces an effective gripping force. Alternatively, the teller or reader can press harder against the money or the page, but this also increases the friction force between adjacent pieces of paper and risks bunching them up. Accomplishing a task efficiently is always about applying just the right amount of force.

While all actions involving contact may ultimately comprise a collection of pushes and pulls, they can occur in confusing combinations that defy easy interpretation or intention. Consider the Chinese finger prison, now known more politically correctly as a finger trap. I first encountered one when a practical joker handed me a woven bamboo tube about three-quarters of an inch in diameter and four inches long. He instructed me to insert my index fingers, pointing toward each other, into its open ends. I felt my fingers scrape along the inner surface of the tube and my fingertips push against each other when they touched near the middle of the device. When he instructed me to do so I began to pull my fingers out, but before I could separate them very much, the tube began simultaneously to stretch in length and shrink in diameter. The harder I pulled, the tighter the tube gripped my fingers and the associated friction forces kept them from escaping the prison. My fingers were caught in a tug-of-war with themselves. I grew increasingly frustrated with my predicament and paused my

A person's index fingers can be inserted easily into the opposing ends of a tube made of helically braided bamboo strips, but any attempt employing increasing force to pull the fingers apart is ill-advised, for it will cause the tube to shrink in diameter as it grows in length, thereby trapping the fingers more securely inside. *Chuck Rausin/ Shutterstock.com.*

prison break. As I relaxed my hands and let my fingers seek each other once again I was no longer pulling on the tube, and so it no longer pulled back. It shrunk a bit in length and grew enough in diameter for me to ease my fingers out of it. In other words, the trick is not to force the issue.

In the typical finger trap, two differently colored bamboo strips are used to create a decorative pattern in the helically woven tube. With one strip running clockwise and the other counterclockwise around the tube, the nature of its construction is evident and helps a prisoner visualize how it responds to forces applied to it. A bamboo strip bends easily enough to be woven, but it can be as strong and stiff as steel and not stretch easily. In the act of pulling the fingers from the tube, each of the bamboo strips is naturally also pulled, but since they cannot stretch individually, the tube must find a different way to match the growing distance of the fingers from each other. It does this by increasing the longitudinal distance over which each strip of bamboo extends, and this is necessarily accompanied by the entire tube shrinking in diameter and so squeezing the fingers with a greater force and trapping them with friction.

The same mechanical principle is employed in a medical device designed to realign dislocated thumb bones, which may be thought of as a pair of misaligned fingers meeting at the knuckle. The device works by the patient inserting the wounded thumb into the open end of a woven sleeve whose other end is tied firmly to an immovable object. As the patient tries to pull the injured thumb out of the sleeve, it tightens its grip, compressing the skin around the separating bones and developing a sufficiently large friction force to pull on the thumb joint until it snaps back into place, allowing the bones to be realigned. It is an ingenious way to treat an injury noninvasively, and all it requires is an understanding of the forces that the patient can feel working.

In the course of walking, every time I touch one of my feet to the pathway I create a pair of matching Newtonian forces. The force of my foot pushing down on the ground may be thought of as the action, and the force of the ground pushing up the reaction. A similar observation may be made about a bouncing ball. Such situations attracted the attention of Heinrich Hertz, the German engineer-physicist whose early interests and accomplishments were in the field of electromagnetic waves. Later in his short career (he died at age thirty-six), he made seminal advances in understanding the intensities of forces associated with elastic bodies pressed against each other. These came to be known as Hertzian contact stresses. According to Hertz, the forces of action and reaction are "reciprocal," and "we are free to consider either of them as the force or the counterforce." Yet because we view the world through anthropocentric eyes, it is natural for us to see ourselves as the main character in most scenes we find ourselves acting in on the world stage. Of course, if some intruder, such as a meteorite, were to strike us, we could hardly deny it the role of actor rather than reactor.

In the absence of a meteorite, if I walk on a sandy beach or a wet field I will leave behind a visible record of my progress, the depressions being evidence that the forces of my feet on the ground were real and sufficiently great to imprint the surface. After a winter snow, I can follow the tracks of animals that inhabit nearby woods, the impressions of their feet and paws being a fleeting record that forces were involved. The spacing between prints tells whether the animal was walking or running. Fossilized footprints can provide a permanent record of an extinct species in motion. Were it not for the force of a foot pressing down on the ground being greater than the ground could bear without yielding some before pushing back, none of these records would exist.

The faster we move, the greater the impact forces our feet (and other parts of our body) feel, because forces applied in motion have an intensifying factor, which engineers refer to as a dynamic effect. This can be demonstrated with a board laid across a ditch. If we walk slowly onto the board, it will bend (and maybe moan) a bit under our weight; if it is strong enough, it will not break, and we will safely reach the other side. If, however, we plod heavily or walk to the middle of the board and begin to jump up and down, it may very well crack and break. The same amount of matter can produce a different force (and effect) depending on how slowly or quickly it is applied and what is the associated acceleration of the contacting object. Fast-moving forces can obviously do more damage than slow-going ones.

This is the reason we are advised to walk softly and slowly across a frozen pond. When we sense a paucity of friction, we tend to take precautions, such as taking smaller steps and placing our feet more deliberately than we normally do. Many other everyday activities depend on the existence of a friction force. As with a lot of things in life, its importance can become most evident in its absence. A pencil will not write very well on a glossy sheet of paper, because there is little or no friction between the point and the plane—no effective hilltops or deep valleys to erode and catch the graphite and leave it as marks on paper. When we make a mistake, we reach for an eraser made of real or synthetic rubber that can produce sufficiently high friction forces to push and pull pieces of graphite out of the crevices as the eraser is alternately pushed and pulled over the error. Pushing harder on an eraser will induce a greater friction force, which might clear the error more quickly but might also catch and push the paper itself to its tearing point. To be effective, forces have to be kept within reasonable bounds.

Ballpoint pens also tend to write better on rougher paper, because the point rolls rather than slips across it. For any round object to truly roll there must be sufficient friction between it and the surface over which it moves, because the act of rolling means that there is no slipping across the area of contact between surfaces. Automobile tires spin in snow and on ice for precisely this reason: there is simply not a sufficiently large friction force available to lift the tire over the rim of even a small depression in which it is stuck. Nothing round—neither tires nor paint rollers nor ballpoints—can truly roll unless there is sufficient friction to overcome slipping and sliding. As with walking, there must be some roughness between the contact surfaces to insure good performance.

But let's resume our thinking about walking. Wearing leather-soled shoes on dry pavement can have the same effect as a bare foot on glazed tile. Boat shoes have textured rubber soles precisely to guarantee a large coefficient of friction between them and wooden boat decks and docks, even when wet, as they often are. The shoes that basketball players wear are designed to exploit the friction force acting on their rubber soles to enable rapid turns, fast breaks, and sudden stops. In executing these moves, a player often experiences a degree of slipping between sneaker and court, which manifests itself in the squeaking that accompanies a competitive game. But too much friction between sole and floor can be dangerous and lead to players tripping over their own feet.

Four-legged animals naturally walk differently than humans, but the forces they experience are very much the same. An adventure of one family's house cat emphasizes the importance of friction to walking by any kind of locomotive creature. Holly was a four-year-old tortoiseshell accustomed to living indoors. One November, her owners took her with them in their recreational vehicle to a rally in Daytona Beach, Florida, a distance of about two hundred miles north of their home in West Palm Beach. Holly wandered away from the RV camp one evening, and she apparently got lost among the three thousand similar-looking motor homes and campers that were parked around the Daytona Speedway. (A fireworks show going on at the time may have contributed to her confusion.) Her owners stuck around and looked for her for two weeks, but eventually they had to abandon their search and go home without her. Two months later, their emaciated and dehydrated cat turned up back in the West Palm Beach area, where a veterinarian read the embedded microchip identifying her owners, who were ecstatic to have her back.

There is no known scientific explanation for how cats can do what Holly did. One thing that seems certain, though, is that Holly did a lot of walking to get back home. When she did get there, her paw pads were bleeding and her claws were abraded unevenly. The front claws were as sharp as ever, but her back ones were "worn down to nothing," according to one report. This is easily explained by the way cats propel themselves forward when walking: they push off with their back feet, the front being used to maintain balance. The worn-down condition of Holly's hind claws was an indication that she had walked a great distance on road pavements or other hard surfaces that ground down the claws the way a file does fingernails— by the force of friction. As a cat exploits the force to push backward with its rear paws and claws on the ground, the reaction from the ground on the claws shortens them a little bit with each pace. This would not happen on a perfectly smooth and slick (and unwalkable on) surface, of course, or even on grass, but it would on the interstate highway shoulder, concrete sidewalks, and gravel paths that Holly likely encountered on her way home. When we bipedal humans walk on such surfaces we usually have our feet protected by shoes, whose soles and heels get worn down by a similar mechanism.

5
Fractious Forces
Shaking and Sliding

The frequency at which a playground swing moves back and forth does not depend on whether a child is riding it or how much the child might weigh. Neither does that of a hypnotist's pocket watch depend on what time it is. It is a consequence of Newton's second law of motion that the frequency of oscillation of any simple pendulum depends only on its length and the local pull of gravity. If the hypnotist were on the Moon, his watch would swing back and forth at a frequency only about 40 percent of what it did on Earth. Every physical system, no matter how simple or complex, has a natural frequency at which it will oscillate when allowed to move freely. This is true of an heirloom watch at the end of a gold chain, an empty rocking chair, and a tuning fork. A sound wave can comprise multiple frequencies associated with the vibrations emanating from the head of a drum, the bars of a xylophone, the body of a bell, and a come-and-get-it ranch house triangle. Some modes of vibration are readily visible, as they are when the wind disturbs a telephone wire, pine tree, or too-flexible bridge.

We may not think much about how our bodies behave when we are standing, walking, running, or jumping in some familiar setting. We have mastered those activities and can perform them while thinking about other ways to move, which we may have to do in a dance, exercise, swimming, or martial arts class. But engineers who give insufficient thought to the natural frequency of the bridge they are designing and how people using the structure may interact with it can put users at risk and the engineers' reputation in peril.

The rhythmic sound of marching soldiers should be a reminder that their synchronized steps impart an amplified force to the surface over which they tread. If they are in step with the natural frequency of a bridge deck, it may be moved to the brink of destruction. Experience of this happening in the past is why London's Albert Bridge—nicknamed the Trembling Lady for the way it shakes under crowds—still bears a sign warning troops to break step when crossing it. For a similar reason, the catwalks erected between the towers of the Brooklyn Bridge when it was under construction were posted with a warning that instructed workers and visitors not to "run, jump, or trot." The caveat to also "Break step!" was signed by chief engineer Washington Roebling, who had learned a lot about force and motion from his father, John Roebling, the designer and builder of several successful suspension bridges before conceiving of the Brooklyn.

The London Millennium Bridge was designed by a committee comprising the engineering firm Arup, the architects Foster and Partners, and the sculptor Sir Anthony Caro. The architects and sculptor understandably emphasized the aesthetics of the pedestrian span connecting the tourist sites of historic Saint Paul's Cathedral and the Tate Modern, the art museum housed in the former Bankside Power Station across the River Thames. With an emphasis on the aesthetics that produced such a striking, extraordinarily low-slung suspension bridge, the engineers involved might have been especially alert to any unusual forces to which this uncommon embodiment of the genre might be subjected.

Engineers involved in the design of the Millennium Bridge made sure that the frequency of up-and-down forces exerted by people walking or running across it in synchrony did not correspond to a natural frequency of the structure. If it did, the deck would begin to rise and fall in time with the footfalls and the bridge's movement could grow in amplitude to a dangerous degree.

But what engineers did not take into account was that when we walk we not only push down and backward with our feet—we also push sideways to keep our balance as we shift from one foot to the other. This sideways motion is exaggerated in the sport of speed skating, where the

The low-slung London Millennium Bridge, which carries pedestrians over the River Thames between Saint Paul's Cathedral and the Tate Modern Art Gallery, had to be closed when forces associated with footsteps caused it to sway excessively. *Image © Daniel Imade/ARUP.*

skaters push their skates at an exaggerated angle to the direction of progress and swing their pendulous arms in great arcs to maintain their balance. Under normal circumstances, the subtle sideways push of shoes on a bridge deck is of little consequence, but in the case of the Millennium Bridge, in which the pathway was suspended as if it were a pendulum, it was crucial because the natural frequency of its sideways motion happened to be very close to that of the sideways force exerted by walkers on its deck. When the bridge began to sway from side to side, even slightly, the walkers felt it and unconsciously began to move in time with the sway. This increased the sideways forces they imparted to the deck and so caused more people instinctively to get in step with the movement, which gave the bridge a collective timed push that caused it to sway even more. According to Michael McCann, an engineer with a special interest in dynamic systems who happened to be on the bridge on a day this occurred, "The people were not consciously marching in step but just retaining balance." He likened the experience to "walking fore and aft on a boat that rolls."

Parents pushing a child in a swing exploit the same phenomenon of re-inforcement when they consciously time each push to coincide with the natural frequency of the swing. It evidently did not occur to the engineers of the Millennium Bridge that they had inadvertently designed a bridge with a natural frequency of sideways motion that was too close to the fre-quency of sideways force pedestrians could impart to it. When the bridge surprised everyone with its resonant behavior, it was promptly closed to Londoners and tourists alike.

After engineers carried out experiments on the misbehaving bridge, they retrofitted it with bracing to change its natural frequency. This al-teration of the structural system necessarily affected the artful lines of the original design. These changes were visible mainly from boats passing beneath, but now aesthetics mattered less than safety. The bridge was also fitted with damping devices, which act like vehicle shock absorbers to keep unwanted movement in check. Some of these devices are quite visible on the retrofitted structure, but given the very public embarrassment of Old Wobbly, as the nickname-prone British had christened the shaky bridge, there could be no practical objection to the changes that were made.

What engineers most wanted to avoid was people on the wobbly bridge being injured by losing their balance and falling. Even when the road does not move beneath us, we can feel uncomfortable at best when walking on surfaces that are not smooth and level. One of our neighbors, a woman in her seventies, broke her wrist when she tried to check her fall after tripping on the edge of a section of pavement that tree roots had caused to be un-expectedly higher than the preceding section. Of course, breaking bones in a fall is nothing new. In 1638, Galileo asked in his *Dialogues Concern-ing Two New Sciences,* perhaps with some hyperbole, "Who does not know that a horse falling from a height of three or four cubits will break his bones, while a dog falling from the same height or a cat from a height of eight or ten cubits will suffer no injury? Equally harmless would be the fall of a grasshopper from a tower or the fall of an ant from the distance of the moon."

A cubit was by tradition a forearm's length (being about twenty inches as measured along my arm from my elbow to my farthest fingertip), and it certainly rings true that a horse is more likely than a cat to break a leg, even if the cat is dropped from a height three times as great. Galileo was basically saying, "The bigger they are, the harder they fall." And, of course, what was true in the Renaissance remains true today. The force of gravity is the same, and the strength of bones has not appreciably changed in the interim. What Galileo did not mention is that all living creatures have an instinct not to fall from even modest heights.

Animals know that the higher the perch from which they jump, the harder they will hit the floor. Ted, our orange tabby cat, liked to roam around our tiered kitchen counter and much preferred to jump down to the floor from its lowest level. He also preferred to land on a soft rug rather than on tile or hardwood. It was not that he feared breaking a leg, for he jumped directly onto the floor from much greater heights, especially when spooked or chased from his perch by our other cat, Leon, a tuxedo. I believe Ted preferred to jump from a lower height because he felt a less jarring force when landing. In other words, all other things being equal, he wished to avoid discomfort. People are the same way: we do not jump from the top of a stepladder but rather climb down from it; gymnasts dismount their apparatus onto a compressible mat. We are at least as sensible as cats when it comes to avoiding unnecessarily large forces that can be painful and even harmful.

Coming straight down onto a rug or mat, as a gymnast does when sticking a landing, means that there will be a perfectly vertical force between mat and feet. There being no horizontal force present, there can be no movement of the mat along the floor. But when our cats jump down onto a rug some distance away, they tend to land at an angle, and the accompanying friction force between paws and nap causes the rug to slide a bit and possibly even bunch up. We humans do the same thing when we walk from a bare floor onto a rug, an action that can push the rug forward and may leave it in some disarray. This can be avoided by placing the rug over a pad or similar gripping device that increases the friction force between it and the floor to keep it in place. On the other hand, a rug pad raises the height of the rug and so creates a tripping hazard. Also, if the pad material disintegrates into a powder, its effectiveness will be canceled. After weighing the risks against the benefits, Catherine and I stopped using rug pads in our house. Consequently, the position and condition of a rug told us a lot about what forces it was being subjected to.

Each morning some years ago, I found a small area rug between my study and the nearest bathroom to be askew. For a while this confused me, because I straightened the rug each night before going to bed. What moved it overnight? I knew that each time I walked over it during the day I rotated it a bit because the doorway to my study is located at a right angle to that to the bathroom, and in walking the short distance to it I necessarily have to make a sharp left turn. Given that the friction between my shoes and the rug is greater than that between the rug and the floor, I impart to the rug a net counterclockwise twisting force. You would think that I might impart a canceling clockwise twist when returning from the bathroom to my study, but the rug had a pile that biased the forces involved. Consequently, I nightly straightened the rug to correct the asymmetry. But what, then, left the rug so out of line the next morning?

Since I do not sleepwalk, I had to look to some other cause. Even if ghosts did exist, they could not be the culprits because they are supposedly massless and so by Newton's law cannot exert any force on what they pass over. Then I thought of Ted. I had seen him race around the backyard, stopping on a dime here and turning on one there. He spent most of his day sleeping, but when startled he would jump up onto all fours and run around the house, making sharp turns with alacrity and agility. He ran from room to room, seemingly to no end but the sheer joy of doing so. When he rounded a turn on the hardwood flooring, the accompanying centripetal force caused him to slide outward on the curve the way a speed skater does during a race. Ted lost his grip because there was insufficient friction between his paw pads and the hardwood to prevent it. On a carpet, when Ted started to run or executed a sharp turn, the friction between paws and pile, not to mention the effectiveness of his rear claws, imparted to the rug pushing and turning forces that caused it to slide and be realigned on the floor. It was on mornings after exceptionally busy nights for Ted that I found the carpet outside my study bunched up against a bookcase and rotated as much as forty-five degrees from where I had left it the night before.

Force and motion go together, with the link between them being mass. Scientists and engineers formalize this connection in what are termed equations of motion, which may be thought of as compact computers. If numbers specifying the mass and acceleration of an object are input to the computer $F = ma$, its output will be the amount of force that caused or resulted from the motion. This kind of knowledge can form the basis for framing and testing hypotheses, as well as developing and interpreting experiments that bring scientists and engineers closer to understanding the sometimes puzzling behavior of phenomena and things both naturally occurring and manmade. Scientists study the former; engineers create the latter. It was by using elemental computers like $F = ma$ that engineers took the first steps in developing rockets that enabled astronauts, satellites, robots, unmanned probes, and rovers to orbit our planet, land on the Moon, rove about Mars, and travel deep into outer space. If there were cats that run and rugs that slide on Jupiter, the same laws of motion would predict how far out of alignment a cat could get a rug.

But we do not have to go to the largest planet in our solar system to witness the movement of rugs on a massive scale. The Sheikh Zayed Grand Mosque in Abu Dhabi is the third largest mosque in the world, capable of accommodating approximately 41,000 worshippers, including more than 7,000 in the main prayer hall, whose sixty-thousand-square-foot floor is covered with a single beautiful hand-knotted Persian carpet. One might think that such an enormous floor covering, which was transported from Iran in pieces that once in place were carefully stitched together, would stay in place. However, the vast number of visitors and worshippers who have walked and prayed on the carpet since its installation in 2007 caused it to move sideways on the floor, opening up gaps between the carpet and the base of columns around which it had been carefully trimmed for a proper fit. To counter this effect somewhat, the locations where tour groups walk and assemble are regularly changed. Even a seemingly immovable object can be moved by a sufficiently large or repetitious force.

6

Lever, Lever, Cantilever

A Single-Handed Force

In high school, my classmates and I learned that many things come in threes. In trigonometry, we were introduced to the properties of figures with three sides and three angles and learned that triangles came in three kinds: right, obtuse, and scalene. In civics, we were instructed that there are three independent branches of the U.S. government: legislative, executive, and judiciary. In Latin class, we translated the opening sentence of Julius Caesar's *Gallic Wars*—*Gallia est omnis divisa in partes tres*—as "All Gaul is divided into three parts." And in physics, we learned about Newton's three laws of motion and that there are three classes of levers, each with three points of interest—the load, the effort, and the fulcrum—with the relative positioning of them distinguishing each class.

According to Thucydides, the ancient ship known as a trireme was developed by the Corinthians in the seventh century BC. It was propelled by a large number of rowers working in groups of three arranged in three tiers, hence the name. The mechanics of operating the oars is the subject of inquiry in the ancient Greek text known as *Mechanica*. It consists of thirty-five numbered problems and their solutions. Although it was once attributed to Aristotle, scholars came to believe with virtual certainty that it was "not the work of Aristotle, though it is probably the product of the Peripatetic School," according to the Loeb Classical Library edition of his minor works.

Regardless of its authorship, the first sentence of problem 3 asks, "Why is it that small forces can move great weights by means of a lever?" The

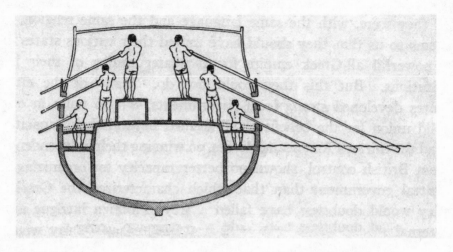

A conceptual arrangement of rowers in a trireme shows how sailors positioned farthest inward from the gunwales worked oars with the greatest lever arm. *From Elson*, Modern Times.

question is not "Can it be done?" but "What is the mechanism whereby it is done?" Likewise, the opening of problem 4 makes clear that in the fourth century BC the operation of a trireme was a question of interest: "Why do the rowers in the middle of the ship contribute most to its movement?" The tentative answer given is that the oar acts like a lever, with the tholepin or oarlock its fulcrum, the water the resistance, and the sailor the force that overcomes it. It was understood at the time that the farther the force from the fulcrum, the greater the effectiveness of the oar. Since the ship is broadest at its middle, the innermost sailors at midships contribute most to its movement.

Because the power of the lever was clearly known, it was reasonable for Archimedes to imagine himself capable of using a very long lever to move the whole Earth, which would continue to be considered the center of the universe until dislodged by Nicolaus Copernicus employing his mental leverage. Illustrations of Archimedes's feat typically fudge exactly on what Archimedes is standing to move the Earth. Almost two millennia after Archimedes, Galileo employed the principle of the lever to explain how a length of timber jutting out from a masonry wall could support a good-size rock hanging from its end but would break if the rock were replaced by a boulder. He began his analysis by assuming that when the beam broke,

An iconic image of Galileo's cantilever beam shows it supporting a
rock, the maximum weight of which
depends on the size of the timber
and how it is oriented. *From Galileo,*
Dialogues Concerning Two New
Sciences.

it would do so at the wall, a reasonable assumption that can be confirmed
by holding a piece of chalk tightly in one hand (representing the immovable wall) and pulling down on the free end with the other hand (the rock).

Galileo viewed the timber beam as a canted (bent) lever—a cantilever—
whose fulcrum he envisioned as being the point where the bottom of the
beam rests on the edge of the wall. Were the beam free to rotate about
this point, the weight of the rock would pull the end down and the top
out from the wall. However, because the beam is firmly embedded, the
cohesive forces holding the wood together would resist such turning. By
equating these opposing effects, Galileo concluded that a long piece of
timber—like a two-by-ten from a modern lumberyard—is capable of supporting a greater weight when the greater cross-sectional dimension is
vertical, the way floor joists and roof rafters are oriented. His analysis provided the basis for the rational approach to structural analysis that is taught
to engineering students to this day. The salient features of what has come
to be known as Galileo's problem are present in an outstretched arm holding a dumbbell. Forces provided by muscles and tendons keep the arm
horizontal.

The lever, in whatever form, is one of the classic simple machines, and
its basic principle is embodied in the seesaw. As we learn in childhood, if
two playmates of about equal weight are sitting at about the same distance

A quarryman can use a crowbar with
a distinctly curved geometry as a
lever that does not require a separate
fulcrum. *From Clker.com, drawing
by Dominique Chappard (@
dominiquechappard).*

from the pivot point, a seesaw will be very nearly in balance. This means
that it takes very little effort for riders to push off from the ground with
their legs to get the seesaw moving and to keep it going up and down. If,
perchance, the two playmates are of noticeably different weights, then
balance can still be achieved if the lighter child sits farther back on the
board while the heavier one sits closer to the center. It is the nature of
levers to work in this way, and scientists and engineers have long understood that balance is achieved when the product of the weight and the
distance of one child from the fulcrum is equal to that of the other. Mul-
tiplying a force times its distance from an axis gives what is known as the
moment of the force about that axis, with the word *moment* relating not
to time but to importance. The greater its moment, the greater the potential of a force to effect rotation.

A crowbar typically has a pronounced curve at its load end, giving it a
hooklike appearance. The shape is not decorative but functional. When
the convex side of the curve rests on the floor close to the edge of a heavy
object, the bar becomes a lever with a self-contained fulcrum. The greater
the curve, the greater the distance the force applied to the crowbar can
be moved down and the higher the edge of the object can be raised. The
curved head of a claw hammer serves much the same purpose: with a nail
in the notch of the claw, the hammer can be rotated about the curved
head-cum-fulcrum by the effort applied at the end of the hammer's handle, thus pulling the nail out. A lever can also take a more abrupt turn by
being shaped like an L. It is the shape of a hand truck used by a delivery
person, with the axle connecting the wheels serving as the fulcrum. It is
how Galileo saw a cantilever beam.

Throughout each day, I use lever after lever. To enter my study, I depress a lever handle to unlatch the door. To turn on the light, I flip a wall
switch that leverages the touch of my finger to make an electrical connec-

tion. To adjust the architect lamp on my desk, I change the angle between its cantilevered arm that holds a bulb and the lever that connects it to the base. Each of these arms is connected to its fulcrum by a spring that holds it in the desired position. The lever is indispensable for converting force into action, and it is unlikely that such a simple machine in some form will ever become obsolete. It behooves us to recognize it for what it is and where it is, however well camouflaged, and to understand its power by feeling the forces it exploits.

I might also have on my desk a can of soda. As recently as the middle of the twentieth century, I would also likely have had nearby a specialized implement to open it. The indispensable implement typically took the form of a four-inch-long by three-quarter-inch-wide by one-eighth-inch-thick piece of steel with one end formed into a shape somewhat resembling a raptor's claw. By hooking a projection on that end under the rim of a beverage can and pulling up on the other end, a user forced the talon-like point into the top of the can, piercing it. Continuing to apply force pushed the talon farther into the top, shearing through it to create a triangular hole from which liquid could flow. The curious implement was clearly a form of lever, and its use provided an excellent means of getting a feel for force and resistance. Since it took the greatest force to initiate the puncturing, using the opener had the familiar feel of getting something started and finding it easier to continue the motion with a noticeably diminished force. Such a "container opener" was invented by DeWitt Sampson of Elmhurst, Illinois, and John Hothersall, of Brooklyn, New York, who patented it in 1935. It was an elementary application of the principle of the lever.

In 1962, the Schlitz Brewing Company advertised an innovative beer can that instead of the conventional steel top had a new aluminum "Sof-top," which could be pierced much more easily. Photographs accompanying an advertisement for it showed a woeful ignorance of and feel for the mechanics of forces and levers. In a photo illustrating the "old hard way," a woman's left hand appeared to have a firm grip on a label-less can and her right wielded the opener with a choked-up grip that would have minimized the mechanical advantage of the lever, thus truly making it difficult to open the generic can. A photo of the "new Schlitz way" showed a can of beer being held in place as daintily as one might hold a tea cup, and

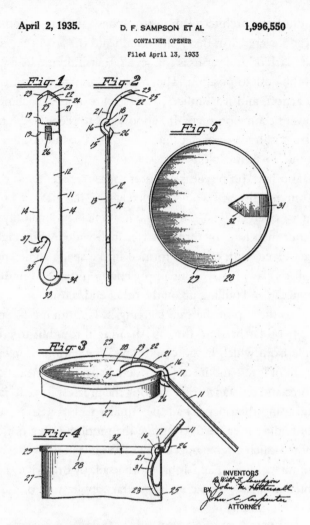

April 2, 1935.   D. F. SAMPSON ET AL   1,996,550

CONTAINER OPENER

Filed April 13, 1933

Patent drawings for a "container opener" show its configuration and illustrate how it acts as a lever in puncturing the steel top of a beverage can. *From U.S. Patent No. 1,996,550.*

the opener held as far from the cutting edge as possible. The ad declared, "Some day all beer cans will open this easy!" Indeed they would, but with a small lever incorporated into the can top itself and designed in such a way that there was no choice but to apply the opening force to the end of the lever.

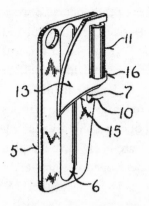
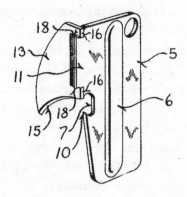

In World War II, soldiers opened their tins of C-rations with a compact device designated a P-38, which was similar to the one illustrated in these patent drawings of a "pocket type can opener" in both its safely closed (*left*) and opened for use (*right*) positions. *From U.S. Patent No. 2,413,528.*

Veterans of World War II remember another kind of can opener, one that was almost as small as the one incorporated into today's pop-top cans. The G.I. P-38 was a one-and-a-half-inch-long, five-eighths-inch wide, and (when folded closed) one-eighth-inch thick implement that became legendary among soldiers and was warmly remembered long after their service. The P-38 is a wonder to hold and behold. Weighing less than a fifth of an ounce, it was light enough to be a barely noticeable addition to the dog-tag chain that all soldiers wore around their neck. Its one moving part was a fold-out knife blade suggestive of a miniature plowshare.

One veteran wrote to me that it was not uncommon to find several P-38s thrown into each case of a dozen cans of food for troops known as C-rations. Directions printed on the piece of paper that some of the openers came wrapped in read, "Twist down to puncture slot in can top inside rim. Cut top by advancing opener with rocking motion. Take small bites." The base part of the little lever had a C-shaped notch cut into its side, which was hooked over the rim around the top of a tin can to provide a fulcrum about which the lever could be worked up and down with one hand as the other held and turned the can in ratchet fashion "to meet the cutting blade," as one patent described its operation. The effort was, of course, applied by the soldier's fingers.

It was said that it took "38 punctures" or bites of the opener to work it all the way around the lid of a Campbell's soup–size C-ration can, and that this is where the little machine got the name P-38. Whether this was the exact number of lever motions required, the finger applying the effort surely felt the force of each one. Another theory is that the name of the device derives from its approximately 38-millimeter length. Some marines called it a "John Wayne," because the actor used one in a military training film. These stories, which veterans love to relate, are likely as apocryphal as the long-held belief that the little opener was developed in thirty-eight days over the summer of 1942 by a Major Thomas Dennehy at the U.S. Army's Subsistence Research Laboratory in Chicago.

According to the website P-38.net, "home of the military can opener," something resembling the P-38 was invented well before World War II. Indeed, the earliest U.S. patent for such a device appears to have been awarded in 1913 to the Frenchman Etienne Darqué for a "tin-box opener" that was collapsible and could be "conveniently placed in the operator's pocket, without danger of injury." The P-38 website lists about ten additional U.S. patents issued between 1922 and 1946 for very similar-looking devices. The curator of the site wondered, "How could . . . these patents have been allowed?" The answer is that although they show only slight changes to Darqué's design, each change was successfully argued to have resulted in an improvement in the performance of the device and how it felt to the operator. For example, the patent issued in 1928 to inventor Dewey Strengberg, of Ishpeming, Michigan, shows the fulcrum cutout to have a ridged surface to give it a more nonslip grip on the bottom edge of the can rim. Chicago inventor Samuel Bloomfield's patent, issued late in 1946, improves on Strengberg's by having a C-shaped fulcrum notch with a sharp point to catch under the can rim. It also had a stiffening groove or "reinforcing rib" down the entire length of the lever.

Just a week after Bloomfield's patent was issued for his "pocket type can opener," another was issued to John Speaker of Milwaukee. One P-38 I have is blind-stamped "US SPEAKER," and it looks exactly like the one illustrated in Speaker's patent. It improves on Strengberg's design by replacing the groove extending the entire length of the lever arm with a trough that stops short of the edges of the arm to which the effort is applied. Because the groove is terminated in this way, the part is less likely to buckle and fold over upon itself and more likely to maintain its shape and be suited for continued use.

It is for similar reasons that structures ranging from aluminum pop-top levers to steel automobile bodies are sculpted rather than left flat. There can be large lessons in such small things, which is why they deserve and repay close scrutiny. By reexamining the simple act of opening a can or bottle—and by feeling the forces involved in doing so—something we might do every day, we can remind ourselves how basic mechanics connects to loftier engineering and design principles.

Ermal Fraze was an engineer who owned a machine tool business in Dayton, Ohio. In 1959, while on a picnic, he found himself thirsting for a beverage, but he remained unslaked because no one had remembered to bring a beverage-can opener. In its absence just the thought of the clever lever made it an implement of torture. Fraze, rather than curse the dryness, resolved to create a can that incorporated an opener into its very design. He succeeded by inventing the pull-tab can, which was the precursor to what became the familiar aluminum pop-top can. Early models of the revolutionary concept had operating instructions imprinted on the lid. If anything is printed there today, it may be a plea to recycle the can or the deposit that will be returned when it is recycled.

Proceeding step by step through the several actions needed to open the familiar pop-top provides a short course in force, resistance, and stiffness, each of which had to be carefully considered in making a can top that will properly and reliably pop. First, with the can held securely in one hand, the index finger of the other must get a purchase on the end of the tab to open a gap between it and the top proper that is large enough to accommodate the fingertip. Second, the finger must pull the end of the tab upward until the other (much shorter) end of the tab touches the metal within the outline incised on the lid, the tab functioning as a lever whose fulcrum is the rivet that fastens it to the lid. Third, the finger must increase its pull on the tab to overcome the top's resistance to being separated along the incised outline; its success will be signaled by a pop and the sound of escaping carbonation. Fourth, the finger proceeds with a lesser force to continue the separation and enlarge the opening to full size. Fifth, the thumb is used to push the tab back over the rivet to allow unimpeded access to the drinking hole. Throughout this process, the active finger and thumb experience only the common forces of push and pull. Generations who have grown up with pop-tops open the cans instinctively. And some do this with the continuous motion of a single finger of a single hand, leaving the other free to click a remote control or reach for a snack. Once opened, today's optimally thin-walled aluminum can is notoriously flimsy and easily crushed by the hand that grasps it. Seasoned users have learned to avoid this happening by holding the can with a Goldilocks amount of force.

Cantilevers have long been ubiquitous. The hoist beams that project from the tops of buildings lining Dutch canals and those that jut out from the openings to haylofts in American barns are cantilevers, as are brackets supporting bookshelves, memory sticks that poke out the side of a laptop, and the broad balconies of apartment buildings. So are airplane wings, sailboat masts, automobile antennas, flagpoles, and chimneys. Parts of our bodies form cantilevers in many familiar activities, including when we reach out to pass a baton in a relay race or to shake hands with a friend or an opponent.

A cafeteria tray carried by one hand is a cantilever, and the food and drink on it constitute the load—Galileo's boulder. Grasping a tray by one of its short sides means it juts out farther from the hand than it does when held by a long side. Even when empty, holding it by a short side demands a greater effort on our part to carry it, because the center of gravity of the tray and its contents is farther from the hand. Adding items of food and drink to the tray adds to the load and hence to the forces required of the hand. The farther away from our grip that we place items, the more our hand will feel their influence—their moment. If the tray is to remain horizontal, the hand must produce a net upward force equal to the weight of the tray and its contents *and* must provide a reactive moment sufficient to keep the tray from rotating downward and spilling its burden. I have observed cafeteria regulars struggle carrying trays loaded with heavy plates placed on them seemingly without any consideration of forces and moments. The forces needed to carry a tray are also affected by how it is held. When two hands grasp it along its two short edges, it is no longer a canti-

A book can be held in one hand as if it were a cantilever beam. In being passed from one person to another, the supporting forces will be transferred from the fingers of the giver to those of the receiver. *iStock.com/Bulat Sylvia.*

lever. If the hands are located at about the middle of each side, it will behave more like a seesaw in the transverse direction.

Analogous forces are present in countless other common situations. Imagine a hardbound book being presented as a prize to a young student. The natural way to hand the book over is front cover up with thumb on top. The fingers underneath will reach farther out along the back cover than the thumb does on top, and the book has become a cantilever with the tips of the fingers acting as the fulcrum and the thumb countering the weight of the book. The mechanics of the arrangement may best be experienced by holding a slim volume in a horizontal position and slowly relaxing the thumb's pressure on the cover. As this is done, the book tilts more and more downward. Clearly, it is just the right amount of downward push from the thumb that keeps the book horizontal. The same holds true for the forces on a dinner plate that is held in one hand as a diner progresses along a buffet line or extends the plate across the counter at a carving station.

The role of the fingers on the back cover is similar to that of the thumb on the front, but they provide an upward push not only to support the weight of the book but also to counter the downward push of the thumb. Together, the thumb and fingers provide just the right combination of forces to prevent the book from tilting or falling to the floor. To get a better sense of the supporting role of the thumb and fingers, we can try to slide the thumb toward a spot just above the fingers. The forces required of both will increase to the point where we can no longer hold the book. The same happens if we move the fingers toward the thumb. If we try to move the fingers to a position directly beneath the thumb, we will find the forces needed to be beyond our ability. This is so because our fingers lack the ability possessed by most gecko species to exert a pull as well as a push on a surface with which they come in contact.

The situation of two parallel forces equal in magnitude but opposite in direction occurs so frequently in the course of the design and use of devices and structures that it has its own name: a *couple*. Daily human activity is full of forces applied as a couple. They might be thought to be akin to identical twins sitting on one of those S-shaped sofas known as a tête-à-tête, each twin staring straight ahead from his or her seat. Just as the twins might have separated themselves from the crowd at a party to sit on this curious piece of furniture, so a mechanical couple can separate itself from any group of parallel forces, leaving the other partygoers to their own devices. This is not to say that couples of force do not contribute to the party, it's just that all they can really do is turn the conversation without moving it. We will open the door and shed some light on this apparent paradox when we discuss the forces involved in turning a doorknob or twisting the top of a bottle or jar.

A couple is also very helpful when we suffer a flat tire. One common way to remove the disabled wheel is to use a four-armed lug wrench shaped in the form of a plus sign. With the socket at the end of one arm engaged on a lug nut, the arms perpendicular to it serve as cantilevers to provide a mechanical advantage in unscrewing the nut. When the nut is screwed on very tightly, the use of two hands is usually required to get it to budge: one hand pressing down on one arm of the wrench as the other pulls up on the opposite one. Together, the forces separated by the combined lengths of the crossbar form a couple that can unscrew the nut without necessarily causing the wheel itself to rotate.

For many people, achieving reading bliss is not fully realizable. In one manageable form, a small to moderate size book can be held open with just three fingers: the thumb pushed into the gutter between the pages to which it is opened, thus keeping them apart; the index finger crooked around the lower back cover and spine, holding the book at an appropriate angle; and the middle finger hooked under the lower spine and front cover, providing a stabilizing force and bearing some of the weight. When we hold a book in this manner, we do feel the forces, which can be tiring.

Many devices are available to assist the reader in holding a book open. An easel is especially favored by cooks because it leaves both hands free. A less commonly encountered implement has the shape of a lozenge with concave sides. It somewhat resembles a kite, but with a central hole into which a thumb can be inserted to push the device into a book's gutter. This causes the sides of the lozenge to press against the facing pages of an open book and keep them in place. In one form it is marketed as a "wooden ring page spreader" with the aid of which one thumb can accomplish the role of several fingers. Among related devices are the weighted strings and leather-covered lead weights used in rare book libraries. Of course, to use these the book must rest on a table or lectern with the reader sitting or standing upright, and reading this way can seem more like work than leisure.

Even when holding a book with both hands, larger volumes can seem to gain weight as our reading progresses, as if turning a page from right to left were somehow adding to the burden. If our reading chair has arms, the book's weight can be transferred to them through the reader's forearms and elbows, but this restricts rearranging the body, which is desirable when reading for long periods. Heavier books can be supported on the lap, of course, but in time the legs want to move, and the weight of a heavier volume may actually affect blood circulation. Some readers prefer a recumbent position, which allows them to rest the book on the chest or abdomen, but this can lead to extreme discomfort when a book's sharply pointed corners dig into the belly.

The electronic book promised to solve a lot of these ergonomic problems. Whether we use a Kindle, Nook, Surface, other tablet computer, smart phone, or similar device, there remains the problem of holding it comfortably for a long while. Most electronic readers have rounded or softened corners and edges, and so they do address somewhat the problem of the thing digging into a reader's flesh. Weight can remain a problem, especially with a full-size tablet. And whatever format an e-book takes, it must still be held with the same hands and fingers that hold print books. And it must still be passed from one reader to another with a form of cantilevered hand action that relies on the coordination of fingers and thumb.

Scissors—effectively a pair of levers linked at a common fulcrum—and other machines with moving parts are obviously more complicated in their construction and operation than eating utensils and other objects of a single rigid part, but they all embody and exploit the same basic mechanical principles. This is why understanding the simple goes a long way toward helping us understand the more complicated. By getting a feel for the forces at play in using the simplest of objects and machines, engineers prepare themselves to identify and understand the forces at work in more complex systems. Great philosophers, natural philosophers, and scientists, from Aristotle to Galileo to Newton to Einstein, seem to have recognized this instinctively. Their greatest works often begin with the simplest—or at least most easily understood—of relevant problems, questions, or thought experiments, thereby serving as starting points not only to encapsulate the subject but also to introduce the reader to a manner of thinking about it. Among excellent examples of this are the stories of two objects dropped from a leaning tower, an apple falling from a tree, and a train pulling out of a station. Scientists and engineers see such stories not as metaphors but as nature in a nutshell from which they can generalize to universal physical hypotheses, theories, and laws.

Whether cutlery or chopsticks, eating utensils held in the hand are effectively cantilever beams. In the United States, the customary way of holding a fork or spoon is remarkably similar to that of holding a dinner plate. All the forces involved are essentially just elemental pushes between skin and instrument, but how those pushes interact with each other makes all the difference in how we perceive them.

Given their instinct to grasp objects, very young children may reflexively hold the handle of a piece of cutlery as one might hold a wooden spoon to stir a tasty stew or a witch's brew. The child's grip also suggests that the utensil might be used as a club to beat the food on the plate into mushy submission. In many social settings, an adult making a fist around the handle of an eating utensil is neither cute nor couth. Squeezing the handle of a spoon gives us little different a sensation than grasping a rope or a baseball bat or any other rodlike object. We can feel that we are compressing our hand around the thing and that it is resisting being crushed, but the wide area across which the forces are distributed does not allow for fine distinctions of what role the individual fingers play.

Very young children may not have the fine motor skills to hold a pencil properly, but in time they learn to do so in an adult way. For me, this means close to the point and within the triad formed by the thumb and first two fingers, letting the shaft rest against the upper part of the index finger for stability and leverage. The forces at play are still essentially pushes— and pulls resulting from pushes. With practice, the pencil becomes an extension of the fingers and the hand and executes loops as easily as lines. As we become more adept with and tolerant of the forces involved, we cease to think about them, if we ever really did. It takes effort to feel the physical force of composition, but ultimately it is what enables us to engage in one of the essentials of civilization—communication through peculiar marks on paper that form letters, words, paragraphs, chapters, books.

I used to write exclusively with wood-cased pencils, but their changing feel as the point wore down was distracting. The point could be revitalized by sharpening, leaving a slightly shorter pencil with an imperceptibly lighter heft and balance. Although the cumulative effect of repeated sharpening did produce an entirely different feeling pencil, it was something that we adjusted to gradually as it evolved. A more abrupt adjustment in applied force is required if we do not want to break the point of a newly sharpened pencil. We will certainly welcome the renewed capacity to draw a fine line, but the lead's brittleness means that we will have to draw with care lest the tip snap off. Only after a little use does a pencil point get burnished to a degree of smoothness and toughness that feels good to me, but that feel is always ephemeral. If I want to continue to have a fine point I have to keep returning to the sharpener. Frustration drove me to use mechanical pencils, whose thin leads maintain a satisfying fineness and, the lead being such a small portion of the whole, whose overall feel does not change with use and does not require a readjustment of the forces by which I hold it.

It is likewise with an eating utensil. Neither a child nor an adult feels much of a change in force on the implement after taking a bite of food off the tines of a fork or sipping soup from the bowl of a spoon. Unless it is made of plastic, the utensil's weight is so much greater than that of the morsel that the center of gravity barely changes. This is why an engineer analyzing the forces between a hand and a fork or, more generally, a wall and a cantilever beam, ignores the morsel on its tines or the weight of the rope suspending the rock. Engineering modeling sufficient to capture the essence of a system of forces is as much about what to leave out as what to include in an analysis.

The system of forces involved in holding even an empty spoon or fork can really be somewhat subtle. In the typical American way, the side of the middle finger serves as a fulcrum about which the force of the thumb pushing down farther up the handle is resisted by the weight of the fork acting at its center of gravity, which is located beyond the fulcrum. The force of the thumb causes the handle of the fork to be pressed against the places of contact with the fingers, and the resulting pressure assures that sufficient friction force is produced to keep the fork from slipping out of the hand. The tip of the index finger usually rests gently against the side of the handle at the ready to press against it and prevent any inadvertent sideways movement. If we think about them, we can feel the forces.

If we hold a fork too firmly for too long a period, our hand and fingers can become tired and our grip uncomfortable. That is why we tend to hold our fork rather loosely, while knowing that if we hold it too casually we run the risk of embarrassing ourselves at the table when it falls from our grip. As we saw in the case of walking or sliding a crate, whenever there is a pressure between two surfaces in contact, trying to slide them relative to each other will be met with some frictional resistance, which in the case of holding a utensil securely is a good thing. Pressing the fork handle against the fingers increases the resistance to its being pulled from the hand; using polished silverware with wet hands will reduce this resistance. Since we generally hold clean forks in dry hands, we come to rely mostly on the pressure the thumb exerts on the handle and whatever consequent pressure and friction forces are available at the fingers to keep it in place.

Without friction, one of the most helpful and ubiquitous of forces on the face of the Earth, the fork would have to be used slowly and deliberately—

or differently—lest it fall from the hand during the scoop, lift, and pivot maneuvers involved in getting food from plate to mouth. Alternatively, if not pulled with sufficient force to clear the lips, the fork and its mound of mashed potatoes could be left cantilevered from the mouth—the bottom front teeth being the fulcrum, the roof of the mouth the effort, and the handle the beam. Either way it is comical and embarrassing but, mechanically, just another cantilever at work and play.

A knife in use is also a cantilever beam, but the means of support is different. Unlike the fork, the knife is properly grasped with a fistlike grip formed with the index finger extended out along the top of the blade and the thumb along the side of the handle, with the remaining three fingers grasping the end of the handle and forming a pocket within which the handle can rest. Friction forces between handle and hand are important to keep the knife from being pulled out of the hand when, for example, cutting a tough piece of meat. As can easily be experienced, simply pushing the knife blade down onto the meat will not cut it; the blade must also be moved in a slicing motion to do the job. The functional purpose of the index finger extended along the top of the blade is to provide a downward force that by the principle of the lever pushes the blade more strongly into the meat to slice it.

Both forces and moments of forces must be maintained in balance at all times if unwanted motions and social faux pas are to be avoided when using cutlery. Engineers call this state of balance equilibrium. But practical engineering is about knowing what is salient information and what is not. Even though we must unbalance forces on one utensil to put it down and pick up another and rebalance forces on it to move from one course to another of a formal dinner, the motions involved are typically rather slow and so any dynamic effects can be ignored. Furthermore, because any morsel on the utensil will likely weigh comparatively little, it too can be ignored in an engineering analysis. Although the forces and motions involved in using an eating utensil may seem more numerous than necessary, for many people they are more easily mastered than knowing which piece of silverware to pick up and when.

7

Forces, Forces Everywhere

Getting Dressed and Going Out

In the kitchen I am provided with opportunities to experience a range of forces. I open the refrigerator with a deliberate pull to break the seal of the magnetic gasket. The handle does not come off in my hand because it is attached to the door with screws that provide a reactive force to oppose my jerk. Inside the refrigerator, everything on the shelves is in the same position it was last night, because the force of gravity, which never sleeps, has kept it from floating around in the darkness. (Had an earthquake occurred, shaking the shelves and everything on them with forces of disruption, I might have found the contents all in a jumble.)

With my left arm, I reach inside the refrigerator for a package of English muffins, then squeeze it with just the right kind and amount of force to pick it up off the shelf and tuck it under my right arm as if it were the bellows of a bagpipe. Also with my left hand, I lift the butter dish from its dedicated space in the door by thrusting my fingers beneath it and placing my thumb over the removable lid and transfer the clamped pair to my right hand. Again with my left, I grasp a jar of strawberry jam by the edges of its lid (expecting it to be screwed on securely) and carry the lot across the room to the table. There, I employ a further variety of forces. I open the package of muffins by twisting the plastic tag bearing an expiration date and take out one muffin, separate it into top and bottom by prying them apart with my opposable thumbs, all the while keeping the whole in place with my other fingers. I drop the halves into the toaster and depress the lever, thus setting the spring that will provide the force to pop them up when they are done.

In the meantime, I lift the top from the butter dish and slice off a pat to place on the hot muffin. Since the jar of jam is new, my hand cupped over the lid has to apply sufficient force to break the vacuum seal. With the jar open, I grasp a spoon and scoop out an amount sufficient to cover the spread of melting butter. Experience has taught me exactly what the right forces to do all this feel like, and my fingers, hands, and arms re-member so my mind need not. The variety of verbs used to describe these actions emphasizes the variety of forces involved in the course of daily life even before I leave for work.

Watching the clock, I lift the muffin to my mouth with just the right amount of force to overcome gravity yet slowly enough to keep the sticky breakfast from slamming me in the nose or flying out of my hand and onto the floor, where Murphy's Law all but insures it will land butter-and-jam-side down. With the muffin only partly in my mouth, I use my incisors with just the right amount of force to shear off a bite. With the aid of my tongue, I push the morsel to the side of my mouth, where my molars chew it into a bolus that peristaltic forces in my esophagus will push into my stomach, where it will be digested by means of chemical forces.

The act of eating is full of sensations of force, which can range from the virtual absence of resistance associated with devouring a melting choco-late truffle to the jaw-tiring exercise of masticating a truly tough piece of meat. Ingesting food and the experience of force go together. There may be no accounting for taste, but there can be a quantitative measure of how we consume everything from the tasteful to the tasteless. We can count the number of chews it takes to eat a truffle (zero) and an obstinate steak. But mostly we enjoy the qualitative pleasures of the taste and feel of food in our mouth. From soft to crunchy to hard, we learn early in our lives to adjust the forces we apply by means of our teeth to deconstruct the food into swallowable bits without breaking our teeth to bits. There may be no more agonizing feeling than to chomp down on a hard nut with too large a force.

We live and can die by forces. The knives on the kitchen counter can be used for good or evil: to carve a turkey or stab a person. Before there were knives, a rock held firmly in the hand could be used to knap flints for hunting or crush a skull for revenge. When force was applied repeatedly through a pounding motion, a stone could be used to mill grain, carve out an obelisk, shape a bronze pot—or it could be used to bludgeon someone to death. How we hold and move a utensil or a tool or a weapon defines the force that results from it; how we direct that force determines what results from its application. We humans have always relied on forces to accomplish both civilized and uncivilized ends, for good and for ill.

After my breakfast of muffin and mulling, I return upstairs to shower and shave. Modern plumbing enables me to enjoy the sensation of being sprayed by soothing warm water driven through pipes by pressure, which is just force spread out over a surface. Too high a pressure released through too fine a nozzle can be painful or fatal: a very powerful and concentrated water jet can cut through steel. Shaving is done with blades so sharp they can easily nick the skin. John Thoreau, Henry David's brother, died within a few days of slicing his finger while stropping a straight razor. The so-called safety razor, which dates from about the time of that incident, exposes less of the blade and requires little skill to use. In the most modern version of the safety razor, the blades are angled in multiple tracks that enable them to be pulled down the cheek with very little risk of nicking and virtually no risk of cutting very deep. The sharp edges of the blades catch beard hair and slice through it with a force that is hardly felt, at least when the blades are new. Ironically, using dull blades can also increase the risk of nicking oneself because the forces needed to maneuver the razor over a lawn of beard are less predictable. Similarly, using fresh blades without adjusting our feel for the reduced forces required can also result in serious bleeding. Getting through daily life unscathed is all about getting the forces just right.

Every physical action I perform involves force. Even when I am just sitting back after breakfast and thinking through the day ahead, my body can be a flock of forces. My heart muscles are contracting to pump blood; my lungs are inflating and deflating as they take in oxygen and push out carbon dioxide. As I think, my eyelids blink and cleanse my corneas the way wiper blades do a windshield. Throughout the day, I will push and pull on doors, steering wheels, computer keyboards, light switches, and buttons on all kinds of electronic devices. We may control some things with our voice, but even then there are mechanical forces involved. Our diaphragm has to compress our lungs, which push air out through our larynx. If amplification is necessary, the sound waves must push on the mechanical diaphragm of a microphone.

Getting dressed requires an array of forces. Clothes are the enemies of efficiency, having as they do the buttons, buckles, loops, hooks, snaps, zippers, laces, and the like that have to be fastened after we have thrust our arms, legs, and feet into shirts, slacks, socks, and shoes. There are at least three or four distinct types of force involved in just putting on a pair of pants, the exact number depending on where we fall on the odd-couple spectrum. Are we closer to a fussy Felix or an awful Oscar? First we lift the pants off a hanger or pick them up off the floor. Then we pull them on one leg at a time, which we can do while sitting on a chair or hopping around the room. There may or may not be a next step, depending on whether we care if our clothes are on straight. If we are like a Felix, we will look in the mirror to check if our creases are sharp and be sure the waistband button is dead center below our belly button. We close the zipper carefully, lest we catch a shirttail in it. Finally, we put on a belt or set of braces, either of which we do carefully so there are no missed loops or reverse twists and the buckle or snap is properly centered. If we are more like Oscar, a mirror will be of no use, because our crumpled pants will not have a crease to serve as a reference for straightness. And who cares if a belt buckle is centered between fly and belly button or a belt or brace is twisted or centered?

During the course of each day I have to take medication at regular intervals. The pills come in a plastic container that is to me neither bottle nor jar. Whatever it is called, it is fitted with a cap that is imprinted with the chew-gum-and-walk instructions of Push Down & Turn. To lift the cap up and off I first simultaneously have to apply a counterintuitive downward push and a twist. The latter requires a couple of friction forces, and the harder I grasp the cap's circumference, the greater the forces I can exert. The knurled surface of the plastic cap consists of a series of ridges, which averts slipping between my fingers and the cap and so lets me produce a greater turning force. This complex force system involved in opening a medicine container may have been devised to keep children from gaining access to adult medicines, but arthritic grandparents sometimes have to enlist their grandchildren to get the push-and-twist caps off.

Taking medicine can involve a surprisingly large number of forces and motions. I once was prescribed some eye drops that came in a small plastic bottle, which itself was packaged, like a Russian doll, inside a childproof container. After I successfully pushed down and twisted off the cap, I next had to twist off the top of the bottle of drops and then squeeze its sides to dispense the drops. And, of course, this had to be done while I was holding the bottle upside down over one eye, which had to be held open with the fingers of my other hand to ensure that the drops fell into the eye and not on the eyelid.

How do we learn to manage all the forces and movements that we do? We certainly are not born with the ability to do so, as we can confirm by observing any infant. A newborn does not have the strength to hold its own head up, let alone the muscular coordination and fine-motor control to feed itself, except for the essential set of mouth muscles that allows the infant to squeeze and suck on a nipple to feed. It takes time for a child to learn just to turn over, to say nothing of crawling and walking. So much of our development and subsequent getting around and getting along involves our ability to produce mechanical actions with mechanical consequences. We remain dependent on others until we master the muscles that control the forces that our bodies need to apply for admission to adulthood. This is what makes muscular dystrophy and cerebral palsy such debilitating diseases, disadvantaging their victims from fully participating in many ordinary activities that healthy children and adults take for granted.

Opening a door equipped with a familiar round doorknob is something that we begin to learn as children. If one of our small hands cannot fully encompass the knob, we can cup it in two. In either case, it is a haptic act aided by an aural cue: we turn the knob until we feel and hear the latch bolt slide free of the strike plate attached to the doorjamb. (Every part of every thing has a name.) When this happens, we change our hand's twisting motion to a pulling or pushing one that swings open the door. The mechanics are similar to those involved in removing the cap from a childproof medicine container; in both cases we must transition among forces that produce turning and pushing and pulling motions. In time, we do it all instinctively, having learned through trial and error what works for us.

In fact, we come to do the most common everyday things so instinctively that we tend to forget that they can be culture dependent. On my first visit to Cambridge University, I followed the directions of my host and duly reported to the porter's lodge in Trumpington Street. There, I was told that I would be staying in the Burrough's Building. To get there I just had to walk across the Old Court, past the chapel, and up the stairs to the front door, the key to which the porter gave me. The directions were simple, and the walk was short. The key fit easily into the door and turned the lock, but no matter how hard I pulled, the door would not open. Perhaps I was given the wrong key, I thought, and so I walked back to check with the porter. That is the proper key, he insisted, and accompanied me back to Burrough's to see what might be wrong. At the door, he inserted the key, turned it, and easily *pushed* the door open! I attributed my embarrassing behavior to jet lag.

For some reason, I had gotten it into my head that in England everything is done opposite to the way it is in the United States. I could even test and confirm this hypothesis by citing how the English drive on the left side of the road; how the titles of their books (at least older ones) read up the spine; and how they eat with their forks upside down. But no number of confirming examples proves a hypothesis, and all it takes is one counterexample to disprove it. My wrongheaded idea was disproven at the front door of the Burrough's Building. As I later learned, it is customary for exterior doors everywhere to open into a house so that the pins that pass through the knuckles of the hinges and about which the leaves rotate are not accessible to a potential intruder. If they were located on the outside, entrance could be gained simply by knocking out the exposed hinge pins, thereby disconnecting the leaves and allowing the entire door to be removed.

Another reason that the front doors to houses do not swing outward is that there may be a screen or storm door present, which would interfere. For the same reason the screen or storm door has to open out. However, at least in colder climates, drifting snow could block a storm door from being opened in an emergency such as a fire, thereby trapping a family behind it. On the other hand, the exit doors of schools, theaters, and other public buildings do swing outward and are fitted with panic bars that, when pushed against by the force of a crowd straining to escape, will unlatch the door. Presumably a forecast of badly drifting snow will have canceled classes and events and so leave the buildings unoccupied anyway.

Doors off public hallways present a still different problem. If a door to a classroom, say, opened out, it could swing into students rushing down the hall to their next class. This is why the entrances to classrooms, especially in newer school buildings, are often fitted with windows or set back in alcoves. Regarding this issue on the domestic front, since a screen or storm door can be seen through readily, there is little chance of a host swinging it into the face of a guest arriving for a dinner party. But this was all too much for a jet-lagged American to think about when trying to gain entrance to a British building in which a bed for him was waiting.

Regardless of which side of a door a knob is encountered on, it can reveal something about the forces that turn it. Over the course of years, the action of so many hands on a doorknob will leave it marked with the telltale signs of having been operated by the abrasive force of friction. A brass knob, for example, will be polished bright around its four and ten o'clock positions (give or take a half hour), where it was most often grabbed by the thumb and fingers, respectively, of a left hand, or the one and seven positions by the fingers and thumb of a right. Many of us work a doorknob by combining the motions of grasp and twist so seamlessly that our hand may begin to twist the knob even as we are closing our grip on it. This has the effect of our fingers slipping somewhat along the circumference of the knob and so polishing an arc of it. The areas of brightness on a knob used by the public signify not the touch pattern of a single individual but a record of everyone who has touched it, attesting that we all operate a doorknob in pretty much the same way.

Sometimes even the greatest grasping pressure we can exert on a doorknob will not turn it. This is the case when there is not sufficient friction between our skin and the knob, perhaps because our hand or the knob is wet or greasy. We also might not be able to grasp the knob firmly enough to turn it if we are aged and arthritic. It is out of such common domestic frustrations, if not downright failures in attempting to operate common equipment, that everyday inventions are born and familiar technology evolves—often along with custom. Typically, early attempts to fix a problem involve retrofitting the existing technology with appurtenances that serve an ad hoc purpose. In many cases, these fixes are devised by individuals for use in their own homes. For example, one way to increase the frictional force between hand and doorknob is to place a tight-fitting rubber band around the knob's periphery. Although not an aesthetically elegant solution, a slip-free rubber band will provide a surface that our fingers can grasp and through it the knob itself. An entrepreneurial homeowner might even think of marketing packages of properly sized and colorized rubber bands as "Doorknob Band-Aides" or some such commercial play on words.

An even less aesthetically pleasing solution might be to wrap the doorknob with something like duct tape or electrician's friction tape. But such solutions cry out for more architecturally integral means of increasing the turning force between hand and knob. Replacing the knobs with ones having ridges or knurls, like those on the cap of a medicine bottle, is another way to increase a hand's ability to exert sufficiently large tangential forces. This is essentially why multifaceted molded-glass doorknobs in older homes are not only attractive but also very effective pieces of hardware.

The problem of not being able to develop enough of a grip (and consequently enough frictional force) between the hand and the doorknob can also be overcome by making a knob not perfectly cylindrical or spherical but rather oblate or prolate, depending on whether it looks like a football in a position to be hiked or to be kicked. This modification makes the knob more ovoid, so it can be turned not by friction but rather by pushing in opposite directions on opposite sides of the elongated knob as if turning a wing nut or working a cruciform water-faucet handle. A doorknob of any continuous shape can be thought of as a continuum of levers, with the spindle being the fulcrum. Applying a force tangential to the knob's periphery is equivalent to applying it perpendicular to a radial line defining each little lever. Given this, it should not be surprising that so many doors are fitted with levers that are explicitly so rather than disguised as knobs. (Where a homeowner does not wish to incur the expense of replacing all of a house's round knobs with linear door opening devices, there are kits that enable a straight lever to be attached to a round knob.) Door levers come in a wide variety of decorative shapes and finishes, but they all enable people whose hands do not have the size, strength, or suppleness to grip a knob to open a door.

Forces and their effectiveness or lack thereof can influence public policy, which in turn can drive technology. In the United States, inventors and manufacturers began to think seriously about door levers as alternatives to knobs in the years leading up to the passage in 1990 of the Americans with Disabilities Act. Like most legislation, the act was preceded by an increasing sensitivity to the problems it addressed. In the 1970s, for example, with a growing public awareness of the plight of the disabled, recipients of federal funds were prohibited from discriminating against people with physical limitations. In the language of the federal guidelines, doors to public buildings had to be equipped with hardware that, according to the U.S. Access Board, has "a shape that is easy to grasp with one hand and does not require tight grasping, tight pinching, or twisting of the wrist to operate. Lever-operated mechanisms, push-type mechanisms, and U-shaped handles are acceptable designs."

The San Francisco inventor Eugene Perry noted in his 1985 patent for a "lever door handle" that "only relatively recently has it been decreed that the handicapped, the infirm and the aged should be provided with the means to conduct substantially normal lives as it relates to public buildings, transportation and thoroughfares." In fact, home improvement stores were increasingly stocking door handles of a somewhat linear design measuring about four inches long from pivot axis to end. Their length and width made them comfortable to grasp and easier to operate. Longer lever handles would make the task even easier, of course, but there would come a point where they would be out of proportion to the persons operating them and also to the mechanical and aesthetic function for which they were intended.

If my recollection is correct, I first noticed four-inch-long door levers—those that retained no vestige of the two-and-a-quarter-inch-diameter doorknob—in hospitals. It may have been intended that these devices be operated with the hand, but they could also be worked by pushing down with an elbow, pulling up with a forearm, or even by being hooked by a pinkie. In any case, the door to a patient's room could be opened without touching its handle with the bare fingers, thereby minimizing the transmission of germs. Hospital environments are full of levers, as anyone who has sat alone in an examination room has had time to observe. Typically, in addition to the jar of tongue depressors and long-sticked swabs, the plumbing is controlled by levers. The sink is fitted not with domestically familiar faucet handles but with elongated paddle-shaped levers, so medical staff washing their hands can turn the water off with an elbow. In some cases, there are no faucet handles at all: the tap is opened and closed with a foot pedal, which is just another kind of lever. The mechanism linking the foot lever to the water faucet valve is often openly visible beneath the sink.

The proliferation of door levers in the American home is explainable not so much by their sanitary benefits as by their practicality and fashionableness. But, as I have written elsewhere, there is no perfect design. As many advantages as the door lever has over the traditional knob, it also has disadvantages. One of the most annoying is the ability of the projecting lever to snag things, such as shirt cuffs, coat sleeves, and pants pockets. This problem can exist even with U-shaped lever designs that turn back toward the spindle and minimize the gap between the lever and the door. As narrow as that gap may be, things will still get caught in it. Indeed, when the gap is very narrow, it can even pinch the fingers of the hand operating the lever. Another and perhaps even greater disadvantage of door levers is, ironically, the ease with which they can be operated. Although children learn soon enough how to operate a doorknob, they can learn even earlier how to operate a door lever. Opening the door may give toddlers access to basement stairways, storage areas, and the outdoors before they appreciate how dangerous such places can be.

When our children were growing up, we lived in houses whose doors had traditional doorknobs. It was not until long after our children had left home that Catherine and I moved into a home whose doors are fitted with levers. This house also has another feature that was new to us—an animal door. It consists of a flap of heavy transparent plastic material whose bottom edge is fitted with a magnetic strip that, in conjunction with a complementary strip on the door frame, holds the flap in a naturally closed position to keep out wind, rain, and, it is hoped, unwelcome guests. The door has no knob or handle; it is designed to be operated by a cat or dog pressing against it, thereby disengaging the magnetic latch and pushing through with its body. When the pet is completely inside or outside, the flap announces this with a click of the magnetic strip signaling the door has returned to its closed position.

It did not take our two cats long to get the hang of operating their dedicated door. Each animal developed his own style of working it: at first, both Ted and Leon were somewhat tentative, pushing slowly against one corner of the flap and then moving their body through with care. Before long, however, they pushed through the door with a single, deliberate motion. When they were trying to get away from a stray dog or another cat, they hit the flap running and reached safety without missing a beat. Ted became a straight-on head-butter at speed; Leon continued to use his paw to break the magnetic seal in one corner before proceeding. After they were in for the night, we usually locked their door by sliding a hard plastic panel behind the flexible flap, effectively blocking access to other cats, small dogs, raccoons, and possums.

That Ted and Leon used the cat door in a slightly different manner was consistent with their each having a different personality. Ted seemed more overtly curious. As a kitten, and even as a teenager, he ran into any open kitchen cabinet and climbed adroitly among the cans, boxes, and bottles inside. Once, when he was still very small, we lost track of him and opened every cabinet door and shone a flashlight inside every closet trying to locate him. We continued to hear his muffled cries but could not figure out where he was. Eventually, by being very quiet and putting our ear to every door in turn, we found him locked in the freezer. To him it was just another cave to explore, and he must have gone inside while the door was momentarily open, for I doubt if he could open the heavy door himself. He certainly was not able to push it open from the inside.

Ted continued to take an open door as an invitation to explore, but in the meantime he also developed another fixation. On those occasions when he wanted in but the cat door was closed, he would jump up and hook his front paws over the lever handle on our back door, pulling it down by hanging from the lever like a gymnast getting ready to begin a set on the uneven bars. When Ted gave up his grip on the lever and fell to the ground, the sudden release of the weight from the handle allowed it to spring back to its horizontal position with a loud clack and a brief reverberation. He effectively knocked on the door to let us know he was there and wished to come inside. This ability to use a noisy by-product of technology served him well, for he was generally a rather taciturn cat and seldom meowed.

The situation was different at the French doors leading out onto our patio. Unlike the windowless back door, these have multiple panes of glass so that we can easily see when there is someone out there wanting in. Ted never hung from the outside levers of those doors; he and Leon simply just stared pathetically through the lowest panes of glass to signal their wish to have us get up from our chairs to let them in. When we did not respond promptly, they pawed and scratched at the glass and wood.

Our French doors are usually locked: the right panel with dead bolts at top and bottom, the left panel by means of a small knob. When that door is locked, its lever handle cannot move and Ted could not hang from it and release it with a bang. When the door is unlocked, all it takes to open it to go out is to depress the lever and pull inward. The cats had seen us do this many times. One evening, after Catherine and I had been sitting and reading elsewhere in the house, we found the patio door ajar. We assumed that we must not have latched and locked it properly, and perhaps a gust of wind blew it open and the cats let themselves out. On another occasion, when we knew we had closed it firmly, we had to look for another explanation for it being open.

We had observed Ted—but never Leon—take matters into his own paws, as he did at the back door. He jumped up and hooked his paws over the patio door's handle, hanging from it. If the door happened to be unlocked, then his weight caused the lever to become depressed, disengaging the latch. When Ted hung from the handle, his back paws executed a climbing motion. When by chance his right rear paw pushed against the secured right French door, his body was pushed back, then his front paws followed and pulled the door ajar enough for him to squeeze through the gap. It was all in keeping with Newton's law governing action and reaction. Over time, Ted learned a further trick. When the lever did not depress under his hanging weight, he hung from it with his right front paw and reached for the small lock knob with his left. It usually did not work, but if he persisted he could not only turn the oblong knob to unlock the latch but also open the door.

By hanging from the lever handle, our cat Ted could unlatch the patio door and let himself out. *Author photos, digitized by Catherine Petroski.*

I first saw him do this one night as I was reading in a nearby chair. Because I had been hoping to catch him in the act, I kept a camera at the ready. I offer here the first pictures I took of Ted breaking out. Subsequently, I did an internet search for photos and videos of "cat opening door" and discovered that Ted's talent was shared by many a feline. Indeed, I found not only that cats can operate door levers but that some also can master perfectly round doorknobs, using two paws the way a child uses two hands to do so.

Every evening, after making sure the cats are inside and the doors locked, I prepare for bed by brushing my teeth. This involves some of the trickiest maneuvers I will have done with my hands all day. Just to get a dab of toothpaste out of the tube and onto my toothbrush involves picking up the tube, unscrewing its cap, setting it down, picking up my brush, positioning it under the opening of the tube, pinching the tube to coax out a dab of paste, replacing the cap, laying down the tube, grasping the brush, moving it back and forth and up and down and around and around in my mouth, turning on the water, moving the brush every which way under the faucet to rinse it off, placing it back in its holder—in other words, combinations of grasping, twisting, squeezing, pinching, pushing, and pulling motions that my hands segue among while I am thinking about none of them. I do not even have to think about how hard I am brushing, because I have learned to do so lightly enough that I do not erode tooth enamel but aggressively enough to dislodge food from between my teeth and around my gums, as well as produce enough friction between bristles and enamel to polish the surface to a tongue-pleasing sheen.

As I walk down the hall to the bedroom, I reflect upon my force-full day: foot forces, leg forces, jaw forces, shoulder forces, arm forces, wrist forces, hand forces, finger forces. I feel that my body has had a workout even without my going to a gym, and my exercise is not yet fully over. I prepare for bed by contorting my body to shed my daytime clothes and don my night ones. I do think of how cats contort their bodies to clean themselves and, if necessary, open doors. If our cats were still with us, they would be curled in tight balls of fur, closed in upon themselves for warmth and comfort. Ted was a light sleeper, Leon a heavy one. Often, by the time I got to bed I found Leon up against, if not on, my pillow. With both my hands cupped against, around, and under his compact body, I gathered him up as if I were moving laundry from the washer to the dryer and repositioned him to another place on the bed, where he would sleep until he woke. When he did that, he would stretch in his supple skin like a jogger preparing for a run. First he would reach forward with his front legs extended well beyond his body, then he would bring them back and arch his back to twice its normal height, then he would begin to walk away from his planted rear paws until they were stretched out so far that his stomach dragged, and finally he would jump off the bed and walk off into the night. Having my half of the bed to myself, I knew I would sleep soundly as I sank down into the cool sheets and soft mattress like a log in a bog.

A few springs ago, Ted and Leon disappeared within days of each other. We announced this on the neighborhood listserv and followed up on every reported sighting. Alas, no matter how much we drove around and how much we called their names, our feline friends were gone. We imagined that skittish Leon may have left because he was frightened off by some loud construction nearby, and his loyal companion Ted followed him. When many months passed without any reported sightings, we gave up on ever seeing them again. After a couple of years, we no longer held much hope that, however far they may have strayed, Ted and Leon would find their way home, the way Holly did in walking from Daytona Beach to West Palm.

Ted had been one of three orange tabbies on our block, so cats resembling him passed across our patio now and then. Though they were the right size and color, none seemed to have quite the same striping or ears or gait that we remembered. None stopped to look through the French doors, which we expected Ted would do. Then we began to find paw prints on the windshields of the cars, the kind Ted used to leave as he climbed onto a roof to gain the high ground and watch down the driveway. One day, Catherine saw an orange tabby walk across the patio and into the carport and huddle under one of the cars. When she knelt down and called softly, "Ted," the cat thumped its tail before walking away. When this kind of encounter happened with increasing frequency, and the cat looked back more deliberately as it walked away more slowly, Catherine and I wondered if it could indeed be Ted.

In time, as this cat crossed the patio he would pause briefly to look in through the patio doors. This behavior began to recur every couple of days, with him gradually approaching closer and spending a longer time staring in. We became increasingly hopeful that our Ted had returned. We never did hear him ask to come inside, but we could see his mouth forming some word before he stepped back and headed toward the carport. Seeing him do this one day, Catherine opened the back door and called, "Ted." The cat meowed audibly, came to the door slowly, entered cautiously, took some food and water gingerly, and asked to leave.

8

Moments of Inertia

Mass Transit and the Transit of Masses

Before the pandemic, the New York City subway system was carrying over five million passengers daily and providing many an opportunity to experience forces associated with commuting. Those between inanimate objects such as a briefcase and a station platform may be insentient, but those communal ones between human bodies crammed into a subway car during rush hour are all too readily felt. When I commuted to work from the far reaches of the city to Lower Manhattan, it was rare to find a vacant seat; I often had to stand for the duration of the ride. In that position, every movement of the bus or subway car was amplified and I had to brace my body against the backward and forward forces of acceleration associated with pulling away from one station and slowing down to stop at the next.

According to Newton's first law of motion, a body at rest will remain so, and one moving at a constant speed in a straight line will continue to do so unless an outside force acts upon it. This tendency to maintain a state of rest or uniform motion is known as inertia. Hertz called the phenomenon a "fundamental law" that may be "inferred from experience."

As long as a subway car is moving at steady speed down a straight section of track, we can stand steady as if we were on solid ground. When the train accelerates, however, the force of friction between the floor of the car and the nonslip soles of the shoes of standing passengers causes them to follow. Yet the rest of the body has no direct contact with the floor, and so it does not feel the same force directly. It wants to remain in place and thus is bent back from the direction of travel. If the car accelerates too quickly, passengers can be thrown against fellow passengers or, if there is no one close by or something to hold onto, off their feet. That is why standing commuters who can do so hang onto straps, bars, poles, and other parts of the car against which they can push or pull to maintain their position during inevitable accelerations.

Even seasoned straphangers, concentrating on the newspaper they are reading, can be taken by surprise by the sideways jerking forces caused when the train shifts from one track to a parallel one and the centrifugal forces that accompany going around a curve. As a commuter, I could not easily see what lay ahead, because the windows looked out onto a mostly dark tunnel wall. But whether or not I could see what was ahead, the forces were capable of throwing my body off balance. The same forces affect all standees, of course, but when they are packed cheek to jowl they cannot, and in fact do not need to, hold onto anything, for they will be supported by the passengers abutting them. They will sway together without falling as long as the last one in line is holding onto or leaning against something.

A crowded bus can provide many of the same experiences, but the passengers can see them coming. My high school was about ten miles from my home, and both were in the densely populated New York borough of Queens. The shortest commute between them involved getting a ride or walking a mile or so to a depot to catch a bus, transfer to another bus, and eventually get left off within walking distance of the school. About twenty of my classmates boarded the same bus, and we competed for space on it with other commuters. In part to mollify them, I believe, the transit authority initiated a special once-a-day express route that terminated right in front of our school. It was very convenient, and because the same group of us rode the same bus every day, we began to scheme about ways to delay the bus's arrival at school. With the excuse of transportation problems en route, we would not get detentions for missing a class or two, our ultimate goal.

The buses assigned to our route were seldom new, and the regular driver was not a disciplinarian. The combination led to mischief involving forces within everyone's control. Among the problems with the buses was their very soft suspension, and it soon became obvious to our gang that a bus would noticeably lean to the outside of any turn, which the driver seemed to enjoy taking at speed. Whenever he did so, we students were thrown toward the outside of the turn in part by the centrifugal force and in part by the lean of the bus. Whenever this happened we cheered the driver and egged him on to do it again. We sat in the back of the bus the way we sat in the back of the classroom, but our collective and concentrated weight was not alone sufficient to cause the tilt to become excessive. The body of the bus neither rubbed against a tire nor scraped the ground. Still, we wondered, how close to tipping that far could we come?

We were not all future scientists or engineers, but everyone soon got a feel for the forces at play and what we could do to amplify them. We figured that crowding to one side of the bus would double the load on that side's suspension system, so we tried that. The first few times we did it, nothing more than a slightly greater tilt resulted, but as we learned to crowd more compactly, we found that we could make the bus scrape the ground and its tires scrape the wheel well. And when half of us charged across the aisle in anticipation of a turn, we could feel the suspension bottom out. Finally, one morning, we heard and felt something in the chassis snap. The bus was permanently disabled, and we cheered at our victory over the machine, but it was Pyrrhic. We had to wait for a replacement bus, which made us so late to school that our antics were beyond suspicious: they were declared vandalistic by the transit authority and truant by the school. We had to serve a week of detentions and find our own way home during the evening rush. As much as we had learned about the forces that could bring a bus to its knees, we received a greater lesson in embarrassment and in responsibility to our infrastructure.

Most commuters in a large city work in tall buildings, in which they rely on vertical transportation systems known as elevators. Although these have a safety feature that does not allow too many people to squeeze into a car, they can still be crowded. However, because there is no sideways motion, the passengers do not have to hold onto anything—except perhaps their stomach, because it will bottom out when the elevator accelerates up and float when the elevator slows down too quickly. But some elevators can still hold surprises.

Some years ago, on a family trip to Las Vegas, we stayed in the Luxor

Hotel, which is the one on the Strip built in the form of an Egyptian pyramid. The geometry of the thirty-story structure and the arrangement of
its 2,500 guest rooms just inside its four sloped faces made it natural for
elevators to be located near the corners of the building. Understandably,
the elevator shafts there are not vertical but follow the thirty-nine-degree
inclination of the pyramid's sides. These "inclinators" are part of the tourist experience, but the hotel's geometrical oddities were not high in our
consciousness when we arrived late at night, after flying through a winter
storm. The inside of the elevator car we boarded was not remarkable; it
had the usual horizontal floor and ceiling connected by vertical walls, just
like so many elevators in which we had ridden. But when the car began
to ascend, the sensation of movement was totally unfamiliar and we were
thrown off balance. Because the car was moving at an angle to the vertical, its acceleration was askew. The ride seemed a combination of one in a
standard elevator simultaneous with one on an airport people mover. Although we were momentarily unsettled, we quickly got used to the unusual mix of forces; it gave us a new appreciation for the idea that things
are not always what they seem at first to be.

What my family and I experienced in Las Vegas may have been akin to
what some residents of Eugene, Oregon, did in 1965. Unsuspecting people who had accepted an offer of a free eye exam were told to report to a
new optical research center. After being checked in, the unwitting subjects
were led into a windowless examination room that was in fact a motion
simulator—a box mounted on a mechanism controlled by hydraulic actu-
ators capable of swaying the room in predetermined ways. The research
subjects were instructed to stand at a mark on the floor and estimate the
height of triangles projected on the wall. As they were performing the assigned task, unannounced to them the room began to move, and as the
motion increased, some participants sensed it and said so; those who did
not were queried about how they felt. The data collected was used to estimate human tolerance for the movement of a supertall building swaying
in the wind.

The experiment was devised by Leslie Robertson, structural engineer
of the twin towers of the New York World Trade Center. They were going
to be the two tallest buildings in the world, but because of their height and
efficient use of steel, each was also going to be somewhat flexible. The

question was: How flexible could they be? The Oregon experiments concluded that 10 percent of people could be expected to notice two to four inches of sway, and the average person about five inches. Keeping the sway of the actual building within bounds would bother the fewest people. It may not have taken an experiment to come to such a conclusion qualitatively, but Robertson quantified what those bounds were. Even though the towers as built could actually sway as much as three feet at the top, damping devices incorporated into the structural design minimized the discomfort of their occupants.

Approximately fifty thousand people worked in the Twin Towers, which hosted another two hundred thousand business and tourist visitors daily. At any given workday hour, the total population might have approached the capacity of some of the world's largest sports arenas. In those cases, spectators seated closest to the playing field may be able to focus on the details of a play, but the occupants of the uppermost seats will have the better feel for the overall action of the game and the flow of the offense and defense. So it is in a supertall building: occupants on the lowest floors may see what wind can do to individual leaves and pieces of litter out on the sidewalk and feel its force as they walk among the swirls, but it is the people on the topmost floors who can see a gathering storm and feel what the force of a hurricane can do to the building itself.

There are situations when a large number of spectators all watching the same sporting event will act in unison. They do so in Cameron Indoor Stadium, where the Duke Blue Devils play their home basketball games. It has a capacity of about nine thousand fans, virtually all of whom are aficionados of the game. The crowd noise naturally varies with the action on the court. As a game progresses in the usual manner, cheers and jeers come and go. When there is a spectacular play—like a monster dunk—the crowd erupts in unison but then settles down like the damped motion of a skyscraper. When there is a lull in the action there tends also to be a lull in the crowd—until a mascot or some other unifying presence wakes them up. In Cameron, for a long time it was a superfan known as Crazy Towel Guy. Season ticket holders, who constituted by far the bulk of the crowd, knew where he sat and, when the team needed a boost, would prompt him to stand up and twirl his towel to incite the fans to get back into the game. The influence of a single individual on the many is a common phe-

A desk model of Newton's cradle can be used for amusement and stress relief, as well as to demonstrate the interrelationship between force and motion. *CoraMax/ Shutterstock.com.*

nomenon. Some football stadiums can hold as many as a hundred thousand fans. It may take only one or a small group of them standing up and sitting down to initiate a ripple that can grow into a wave progressing around the entire arena. The initiation and continuation of motion can be hypnotic and can catch in its progression the most hardened of spectators.

A related phenomenon is epitomized in a device known as Newton's cradle, which is often marketed as an executive desk toy. One model consists of a handful of steel balls suspended by lengths of fishing line from a chrome-plated frame. Sitting motionless on my desk, the five pendulums evoke the image of a line of straphangers in a subway car. If I nudge the frame sideways, they will respond like passengers leaning in unison in response to the force that accompanies the car's leaving from or arriving at a station.

If one of the end balls is raised to the side and released, the simile ends and the fun begins: the single swinging ball arcs toward the line of stationary ones, which stop it abruptly. A rapid series of steely clicks follows, during which time the hanging balls appear to remain stationary until the ball on the other end of the line swings up. After reaching its apogee, that ball swings back down to strike with a click the four waiting balls. As before, the moving ball stops abruptly, the three middle balls click in place, and the ball that started it all swings up to a position imperceptibly different from where it was when released. And so the sequence repeats with an audible click accompanying the impact of ball upon ball, the sound stopping momentarily as the last ball swings away from the crowd and then returns to it, with a fresh click. The intermittent repetitive motion is hypnotic and can provide a transformative experience of the effects of force made visible and audible. The successive ball-on-ball impact involves our human senses in a crisp display of movement driven by contact, which has been described as "the most fundamental phenomenon in mechanics."

In the seventeenth century, the natural philosopher Isaac Newton formulated the eponymous laws that explain the relationship between force and motion exhibited in the toy. One implication of Newton's second law, $F = ma$, is that when a mass is in free fall its acceleration is that due to gravity, which is customarily denoted g. In this case the equation tells us that weight W and mass m are related through $W = mg$, which explains why the same mass will have a different weight and behave differently on planets with different gravitational pulls.

If there is no force at all applied to an object, it cannot accelerate, which in turn means that it must remain stationary or move in a straight line with no acceleration, which is a state of constant velocity. A mass in motion possesses momentum, which is equal to the product of its mass m and velocity v. When the ball that was raised is just about to swing into the closest of the other four balls, it has a momentum mv, whereas the stationary balls have zero momentum. During the collision, the first ball transfers its momentum to the ball it hits, sending it into the next ball, and so on until the next-to-last ball in the line transfers its momentum to the last, imparting to it a velocity equal to that which started the chain reaction. Since this ball has no other in its way, its momentum continues the motion in an upward arc, until it is checked by the force of its own weight. The ball then retraces the same arc on its way back down, whereupon it strikes the line of four waiting balls, and the motion repeats.

An ideal assemblage of balls must also exhibit the physical law of conservation of energy, which means that the total energy of the system must remain constant. Mechanical energy in such a system consists of two kinds: potential energy, which depends on the ball's weight and the height at which it is positioned; and kinetic energy, which depends upon its mass and the square of its velocity. When all five balls are sitting still, their total energy is zero. When the first ball is raised it gains a potential energy that will be transferred in the form of kinetic energy to the ball it strikes. Since the three middle balls are constrained by their neighbors, they do not move, but the energy is transmitted through them to drive the last ball away with the same velocity that the first one brought to the ensemble. And the sequence repeats. The back-and-forth action of the balls is suggestive of a cradle being rocked, hence the toy's name.

In the course of their interactions, the balls in Newton's cradle certainly have forces acting between the intermediate ones, but these equal and opposite forces are internal to the system and so are canceling as far as it is concerned. They manifest themselves only in transferring the push, momentum, and energy from one ball to the other, until the last is reached. Were the pendulum bobs to be not steel balls but Nerf balls or eggshells, the toy would not work. The spongy Nerf material would so quickly consume energy in the process of being squashed that the cradle would barely rock; the eggshells would be broken, thereby consuming the mechanical energy of the system. The steel balls, being hard and resilient as they are, allow the forces to act efficiently by bouncing into and off of each other without doing damage and so allowing the motion to proceed. Thus, in addition to the forces involved, the materials determine greatly what will result. In a less-than-ideal system, the balls will eventually stop moving because with each collision the system's mechanical energy will be diminished as some kinetic energy is converted into the sound energy that we hear as the click of impact when ball strikes ball.

Even without appealing to concepts of energy and momentum, a properly made desktop Newton's cradle can give us a good feel for mechanical causes and effects generally. Imagine a simplified cradle consisting of only two steel balls, with the left one having been raised and released toward the right one, which is sitting still. We can imagine the moving ball to represent a car and its driver coming down a hill and the stationary ball to represent a car and its driver stopped at the intersection at the bottom. Imagine further that the driver of the speeding car is texting and so does not see the stopped car and rear-ends it. The collision brings the speeding car to a stop and propels the stopped car forward, usually leaving skid marks that tell how far it traveled. If they are wearing seat belts, the drivers will stop or move with their respective vehicles. The speeding driver will feel the force of the seatbelt restraining the body as inertia sends it forward; the stopped driver will feel the force of the seat back propelling the body forward and, if the headrest is not properly positioned, may suffer whiplash. Without a seatbelt, the driver coming down the hill is likely to be thrown into an airbag or through the windshield of the car; the fate of the person in the car struck from behind will depend primarily on the efficacy of the seat back and headrest. Newton's cradle can give us a feel for the mechanics of impact, but in the real world where people are not as hard as steel balls, the consequences can be deadly.

Analogous actions and reactions occur in sports, so many of which involve one object or player striking another. This occurs on a pool table when the force of contact of the cue ball causes the ten numbered balls in their triangular arrangement to bang against one another before breaking out in all directions, like the finale of a fireworks show. Similarly, a bat hits a baseball, a foot kicks a football, a racket swats a tennis ball. In all cases, the nature of the ball and of the thing that strikes it, along with how squarely they interact, determines the subsequent motion. Basketball players increase their chances of a shot going into the basket by giving it some spin, which affects how it ricochets off the backboard and bounces off the rim. Soccer players employ a banana kick to curve the ball around someone or something between them and the goal. Practiced pocket billiards players are notorious for the English they impart to a cue ball to make it jump or curl around one or more balls on its way to striking its target. All such tricks exploit the nature of how an object responds to the forces and moments applied to it. Whether or not they are fluent in Newtonian mechanics, those who perform such tricks have developed a feel for the game and the forces involved.

The performance of sports equipment is likewise heavily dependent upon how it transmits or absorbs the force of what it hits or is hit by. In baseball, the catcher's mitt is heavily padded, enabling its wearer's hand to survive the impact of pitched balls traveling at speeds approaching one hundred miles per hour. The first baseman's glove is not as heavily padded, but its pocket is elongated, so that the player can stretch and reach for the ball thrown from across the infield without putting his fingers at risk or pulling his foot off the bag. Likewise, football players wear helmets, shoulder pads, and other protective equipment to absorb the impacts of the game. Football receivers wear gloves that are coated with silicone or a similar no-slip material or are made of material that is itself high-grip, which helps them make spectacular one-hand catches. The texture of such gloves somewhat mimics the wrinkles that develop on our fingertips when they are soaked in water. Biologists have found that the wrinkles improve our ability to grip wet objects and thereby may have provided an evolutionary advantage to our distant ancestors.

A batter's performance is no less dependent upon the nature of his equipment and how he wields it at home plate. Major leaguers have used wooden bats exclusively, favoring hardwoods such as ash and sometimes maple. Hickory was used in Babe Ruth's day, but it is denser and so is heavier and more difficult to swing. The illegal practice of hollowing out a bat and filling it with cork does lighten it and render it easier to swing, but the reduction in weight negatively affects the velocity of the ball as it leaves the bat, effectively canceling out the advantage gained from a quicker bat speed. No matter of what it is made, a baseball bat will make a characteristic sound when it strikes the ball, and the nature of the sound can telegraph to the fielders and runners on base whether the ball will be going to the infield or outfield or farther.

The act of hitting a baseball may be analogous to one ball hitting another in Newton's cradle, but the elongated nature of the bat makes the situation less straightforward. Depending on how the hitter is holding the bat and where along its length the ball is struck, the player's hands can experience anything from the satisfying feel of hitting a solid home run to the stinging and stunning feel of the bat vibrating uncomfortably. Outside the majors, the hollow aluminum bat has gained considerable popularity. The hollowness makes for a lighter and more flexible bat perceived to impart more energy to a ball.

However initiated, all motion can be broken up into two basic components: translation and rotation. Translation is when the orientation of a moving object always remains parallel to its original position; rotation involves a change in angle. An airplane flying in still air translates in the direction that its fuselage points. If it is moving through a steady crosswind, it will translate in a direction determined by both the forward thrust of its engines and the sideways push of the air. An oscillating rotation of an airplane can occur in a variety of ways: if its nose goes up and down while its tail goes down and up, then it is said to be pitching; if the nose moves left and right and the tail right and left, it is said to be yawing; if the left and right wings alternately move up and down, it is said to be rolling. A plane rotates most notably when it banks to make a turn. The total movement of a plane or any other object is a combination of translation and rotation caused by the forces acting on it, and the attentive passenger may feel the effects of those forces.

Ships are also subject to combinations of force that produce varieties of motion. At 1,300 feet from stem to stern, the cargo ship *Ever Given* was as long as the Empire State Building is tall. In early spring 2021, it was sailing from China to the Netherlands via the Suez Canal, through the narrowest part of which it was transiting on a day with strong crosswinds. The ship's great size was made effectively greater by the shipping containers piled nine high on its deck. This presented a very large surface to the strong crosswinds that were pushing the vessel close to the far bank. This naturally divided the water into two streams and brought into play a phenomenon known as the bank effect: the water in the narrower bankside channel became the faster moving, which lowered the pressure with which it pressed against the hull, which in turn allowed the ship to draw even closer to the bank. The ship progressing through the divided streams of water was in fact akin to a plane's wing slicing through the divided streams of air. Whether it was due to an attempt to counter the bank effect with an adjustment of the rudder or to a lull in the wind, the ship began to yaw uncontrollably and its bow went aground.

Since the length of the ship was almost equal to the width of the canal, it blocked all shipping for about a week, as efforts were made to refloat the vessel. Early attempts failed, but the expectation of an abnormally high spring tide provided a new opportunity, and thirteen powerful tugboats managed to free the ship. Had the maneuver not worked, some of the eighteen thousand shipping containers, each weighing up to forty tons, would have had to be offloaded one at a time in an effort to make the ship lighter and so float higher.

The problem and solution may be explainable in terms of physical force, but the consequences for world commerce were due to economic forces. The Suez Canal is a critical sea route between Asia and Europe; with the canal blocked, the world's supply chain was seriously interrupted. The alternative sea route via the Cape of Good Hope at the southernmost tip of Africa would add a couple of weeks to the journey. Shippers and shipping companies had to make risk-benefit calculations about whether to go that route or lay at anchor until the canal reopened. Among the ships waiting it out were tankers delivering oil to Syria and container ships carrying toilet paper to Europe. Fears of shortages led to rationing as the less tangible forces of supply and demand began to come into effect.

The problem surfaced again in the fall of 2021. This time the choke point was not in a canal but in the ports of destination of the massive container ships that take a long time to unload. With transpacific trade recovering from a pandemic-induced slowdown, so many ships were arriving on the West Coast of the United States that many had to wait at anchor for as many as ten days until earlier arriving ones could be offloaded. At one point, almost seventy-five ships were waiting to enter the ports of Los Angeles and Long Beach, through which pass 40 percent of all shipping containers coming into the United States. Among the supply chains affected was the publishing industry, many of whose more color-intense books are printed in Asia. A publisher might already have incurred a charge of $25,000 to ship a steel box containing thirty-five thousand books across the ocean, and a delay in getting them onto bookstore shelves in time for the holiday season could make the difference between a financially profitable best-seller and a money-losing flop. As important as the physical forces on ships and the containers they carry are, the economic forces can be determining factors in whether an enterprise is a success or a failure.

Problems did not end when big steel boxes by the thousands were repositioned from ship to shore. At one time as many as eighty thousand shipping containers were stacked up at the Port of Savannah, the third largest container destination in the country, leaving little room for new arrivals. To accommodate the volume of boxes they were stacked up to five high, which made it more time consuming to retrieve a specific one to place on a waiting truck. Under normal conditions, containers would be moved away promptly, but the pandemic had created a shortage of truck drivers. Some big boxes were being delivered to warehouses, but space in them was also nearing capacity. Such conditions were wreaking havoc with the global supply chain, and there was not a single problem to zero in on and resolve to fix the entire system. Virtually every link in the chain seemed to be on the verge of being overloaded. It was a situation familiar to engineers who study failure: the total collapse of a bridge or building is seldom attributable to a single cause. Whereas too large a force applied to one beam or column may be the trigger, that element may have been weakened by careless construction, collision, or corrosion to be incapable of bearing the force it was designed to carry.

A Ferris wheel combines translational and rotational motion in a rather pleasant way. The wheel obviously rotates around its elevated axle, driven by the force transmitted to it from an engine or motor by belts, wheels, chains, or gears. The cars on the big wheel may be thought of as attached to its periphery by means of smaller axles. If the wheel rotates slowly at a constant rate, then a freely hanging car will translate in a circle, its floor always remaining parallel to the ground. If the car's occupants find such tame movement too sedate for their liking, they can rock back and forth and get the car rotating about its axle while continuing also to translate in a large circle. The Ferris wheel, with the range of motion available to its cars, can thus be many things to many riders. But when motion is not under our direct control, we can become uncomfortable or even alarmed by it, which is why some people feel more at ease when they are driving a car than when they are riding in the passenger seat.

Other people are happy to give up control. These are ones who in an amusement park eschew the Ferris wheel and merry-go-round for the roller coaster and even more forceful rides. One of these is the misleadingly named Da Vinci's Cradle at the Busch Gardens Theme Park located near Colonial Williamsburg in Virginia. It consists of a large gondola in which ten rows of benches seat four people each. The gondola is supported from a structural frame by large pendulumlike links and is embellished by large pulleys and belts that may or may not be functional. The floor of the gondola remains horizontal throughout the ride, which means that its motion is purely translational. Considerable acceleration forces do develop as its points of support fall and rise, carrying the cradle and each rider in large circular arcs. Because of the centrifugal acceleration associated with such motion, the riders' stomachs experience forces akin to those in a swiftly rising and falling elevator.

Whereas the motion experienced in Da Vinci's Cradle remains in a vertical plane, that experienced in a roller coaster car can involve turns and twists well out of a single plane. This means that the occupants of the cars feel not only the rises and dips that they do in the cradle but also sideways and even upside-down motion that throws their bodies in multiple directions. All amusement park rides are but variations on these themes. The challenge to the designers of these rides is to ensure that the motion of the vehicle is seemingly erratic but always under control. To accomplish this, the ride engineer must ensure that the forces associated with the motions can be tolerated not only by the structure itself but also by the thrill seekers it carries.

As children, we came close to experiencing pure rotational motion when we twirled our bodies round and round until we became dizzy and fell to the ground laughing. With practice we may have been able to learn to spin the way ballet dancers and ice skaters do, but few of us got beyond thinking about how they did it. Since the force between an ice skate and the ice is minimal, once a spin is begun there is little that can be done using the skate alone to speed up or slow down the rotation rate. But skaters can control the speed of the spin by redistributing the mass of their body, which is done most easily by changing the position of the limbs. If while spinning they hold their arms above their head, making their body as axially compact as possible, they will rotate faster; if they stretch their arms out horizontally, they will slow themselves down. In the first instance, they are compacting their mass; in the second, they are dispersing it.

A dancer's free and graceful movements around an ice rink might not seem to have anything in common with a primitive steam engine puffing down a straightaway, but there is a connection. If the engine were run at too high a pressure, its boiler could explode. To limit operational speed and pressure, the engine was fitted with a mechanical device called a centrifugal governor, which consisted of a vertical shaft whose rotation was

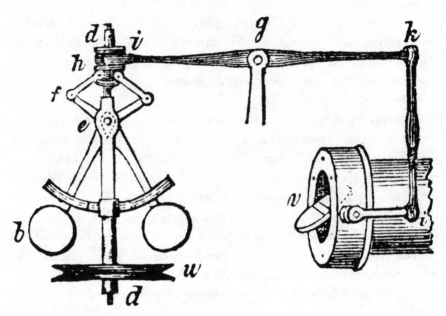

For an old centrifugal ball governor (*left*), any change in the speed of a steam engine to which it was linked (*not shown*) caused a change in the rate of rotation of the ball assembly, which altered its overall configuration and mass distribution. This in turn caused the collar (*h*) to move along the vertical shaft and the linkage to adjust the position of the choke valve (*v*) and so keep the steam pressure and engine speed in check. *Morphart Creation/Shutterstock.com.*

driven through gears by the engine itself. A pair of heavy metal balls linked to the rotating shaft spun around with it and, by the design of the linkage between them, exploited the centrifugal forces that caused the balls to move in a larger or smaller circle as the speed increased or decreased. The linkage was also connected to a valve that opened and closed accordingly, thus regulating the amount of steam reaching the engine's cylinders.

We can easily experience the feel of centrifugal force by twirling above our head an object tethered to the end of a string. The faster the object travels in a circle, the more horizontal its path will be and the greater will be the force in the string and felt by our hand. If the force exceeds the strength of the string, it will break and the object will fly off on a tangent. The same phenomenon can be achieved by releasing our grasp on the string before it breaks. In the track-and-field event of hammer throw, a sixteen-pound weight at the end of a four-foot-long steel cable deliberately let go of by a spinning athlete can travel almost three hundred feet. The South American throwing weapon known as a *boleadora* or bola consists of weights at the end of one or more interconnected cords. When twirled above a gaucho's head and expertly let go, it can entangle the legs of errant cattle and hunted prey alike.

The awkwardly named and potentially confusing concept of moment of inertia is a measure of how the mass of an object is distributed about an axis of rotation. In spite of the similarity of phrase, a moment of inertia is not nearly the same as the moment of a force, which is simply the product of a force and its distance from an axis. It is the moment rather than the force itself that most influences rotational motion. When children rolled hoops, the moment was produced by using the stick to strike a glancing blow along the hoop's rim to keep it rolling down the road. Just as a force accelerates a mass in a straight line according to the law $F = ma$, so the moment of a force accelerates an object's rotational motion. This is where the moment of inertia—designated I—comes into play. An engineer expresses the relationship between moment of force and moment of inertia as $M = I\alpha$, using the lowercase Greek letter alpha to represent angular acceleration—that is, the increase over time of the angular velocity, usually represented by ω (lowercase omega). The concise formula $M = I\alpha$ is clearly analogous to $F = ma$, which is as familiar to engineers as $E = mc^2$ is to just about everyone. If a rotating object has no moment applied to it, then there will be no angular acceleration, and hence its angular velocity will be unchanged, as will its angular momentum, $I\omega$. Since the moment of inertia of a symmetrical object such as a hoop or a wheel does not change as it rotates about its center, the angular velocity will continue at a constant rate until the moment of a friction or some other retarding force slows it down or stops it.

The concept of moment of inertia is central to understanding the behavior of such seemingly disparate things as acrobats, bicycle wheels, and gyroscopes. It also explains the mysterious phenomenon of a long, flat spinning object reversing its direction of spin for no apparent reason. This happens when a tennis racket held with its face parallel to the ground and thrown head over handle into the air comes down with its head oriented perpendicular to the ground. The change in orientation is explained by the fact that every three-dimensional object has three principal axes of rotation, and a moment of inertia is associated with each of them. For a tennis racket, the axes are: (1) along the line of the handle, about which many a player twirls the racket while awaiting a serve; (2) perpendicular to the plane of the racket and through its center of mass; and (3) perpendicular to the first two. If the racket is thrown up with a spin about this last axis, after one rotation the handle will return to the player's hand but with the head facing sideways.

Since this is a strictly mechanical phenomenon, it could have been observed millennia ago for objects with a certain kind of geometry. Such objects have been found in the form of stones by archaeologists interested in prehistoric ax and adz heads. The stones, known as celts, had been eroded into smooth shapes, the way river gravel is by the water current. That an ancient stone and a modern manufactured object manifest the same kind of behavior confirms our understanding that laws of motion are timeless and inherent in nature. After observing the movement of objects in the solar system as well as on Earth, scientists formulated the laws; they did not create them.

A celtic stone made of plastic accompanied a promotional mailing announcing the third edition of a classic textbook on engineering mechanics. The so-called rattleback is about three and three-fourths inches long, three-fourths inches wide, and half an inch deep, and it has a flat top and an ellipsoidal bottom, whose major axis is offset slightly from that of the flat top. This biases the professorial toy to spin in a counterclockwise direction. If its rounded side is set down on a horizontal surface and started spinning in the clockwise direction, it will rattle forward a few times, stop momentarily, and without human intervention begin to spin in its preferred counterclockwise direction. The phenomenon has come to be known as the Dzhanibekov effect, not because Russian cosmonaut Vladimir Dzhanibekov discovered it but because in 1985 he observed and videotaped an example of it in the microgravity of the International Space Station. His video shows a T-handle spinning off a screw thread and reversing its direction of spin without encountering any other object.

The concept of moment of inertia also provides an opportunity to interject a bit of whimsy into what might be considered a tedious analysis of mechanical phenomena. Engineers are not known for their sense of humor, but they do enjoy a bit of levity now and then, even if it tends toward the technical. The typical punch line of an engineer's joke depends upon a certain familiarity with jargon. One such joke has a pair of engineers hunting a lion. The engineers know that the lion is stronger and faster than they, and so they discuss how they might catch him before he sees them trying to snare him. One engineer proposes to the other that they watch the lion from a distance and approach only after he has dozed off and thus catch him in one of his moments of inertia. Such attempts at humor may make some engineers roar with laughter, but others tend neither to get the joke nor laugh at it if they do.

For a child, a bicycle can provide an introductory course in force, motion, and balance. Learning to ride a two-wheeler is difficult for most children because the forces involved are unfamiliar and counterintuitive. A child's (and a nervous parent's) inclination is to begin by pedaling slowly. However, the very forces that stabilize a bicycle do so more effectively at higher speeds. These are the gyroscopic forces that keep the bicycle's front wheel on a straight course and make it easy to ride with no hands. In fact, the hands on the handlebars are more important for steering the bike into a turn than keeping it going in a straight line. It is only when children learn to entrust their safety to the mysterious forces involved that the art of bicycle riding can be mastered—and exceeded, or so it may seem, by doing jumps, wheelies, and other circuslike maneuvers.

When it was more common for young people not only to ride but also to maintain a bicycle, the leg-propelled machine provided a close-at-hand introduction to the vagaries of angular motion. In order to repair a bicycle's flat tire, a kid would turn the bike upside down to rest it on the tripod comprising the ends of its handlebar and its seat. The bolts holding the wheel's axle to the bicycle's fork were loosened and the wheel lifted off. After fixing the flat but before reattaching the wheel, the inquisitive young mechanic often experimented with it. With the wheel held by the ends of its axle, it could be got spinning by scuffing it along the ground. When a wheel is spinning fast, it has rotational inertia, which makes it difficult to realign the axis of rotation while the wheel remains in the plane of its spin. This is the gyroscopic effect, and it can be demonstrated in a physics class by having students in turn sit on a stool mounted on a horizontal turntable and hold a spinning wheel. If they try to tilt the wheel off the vertical, they will feel the couple of forces opposing the reorientation. The resulting torque will be transmitted through the demonstrator to the turntable, which will then begin to rotate.

After a flat tire was fixed and the wheel reattached to the bicycle, the young mechanic would have wanted to be sure it was properly aligned. If an arc of the tire rubbed against the fork, the correction involved selectively tightening individual spokes in the right order, a nontrivial and time-consuming procedure that rewarded its performer with a sense of tension. If the rubbing occurred all around, it could be fixed by repositioning the axle in the slots of the fork. When all was in line and all the nuts fully tightened, it was hard to resist getting the wheel spinning as fast as possible, just for the fun of it. In order to achieve maximum angular rotation and momentum, a hand was usually swiped repeatedly across the spinning wheel so that the friction force between palm and tread imparted a moment about the wheel's axle. Just as much fun as getting the wheel going fast was getting it to stop. This could be done by using a hand as if it were a caliper brake—advisedly with the protective aid of a glove or a rag—grabbing the tire's sides quickly and letting go before the hand was burned or fed into the fender the way a salami is into a meat slicer. With the flat fixed and the bike right side up, the wise mechanic rode it around the block to be sure all was in proper working order.

Whether or not we are conscious of it, we are constantly imparting forces and motions to and receiving their effects back from everything we touch. The variety of such experience enables us to become familiar with the physical world and its ways. When we encounter forces and motions that are outside our library of experience, we are taken by surprise, sometimes to our detriment, sometimes to our benefit, and sometimes simply to our sheer enjoyment.

Exploitation of force as a natural resource does not have to be confined to scientists in a laboratory—on Earth or in space—or engineers in the office or in the field. Everyone can harness force for fun, and thinking we are too much the adult to do so can be harmful to our mental health. All work and no play makes Jack's father a dull dad. How refreshing might it be if when Dad took young Jack and his sister Jill to visit their grandparents they told stories of how they and, later, their father and his friends had fun before there were electronic toys that separated the player from the play, the forcer from the force? They could tell how they rolled hoops with sticks and rolled themselves down hillsides and rode their bikes not to get anywhere but just for the fun of pushing down on pedals to roll forward. They could also tell the grandkids how in their father's time he took tin cans from a grocery bag and just watched them roll across a sloped floor or observed that two cans of identical external geometry reached the baseboard at different times, depending on the contents identified on the label. They might also tell how he and his friends had soup-can races, pitting vegetable against tomato, say. The winner depended not on the taste of the soup but on its consistency in the can and, ultimately, on the moment of inertia of the can and its contents.

Who can see or feel the virtual forces and torques being simulated in an electronic game? If the feel of a joystick or action button is invariant, it cannot convey the sense of control that we experience holding the handlebars of a bike or giving a basketball backspin to make it return to us after hitting the ground. It does not provide the shock that a bowling shoe does when it stops us abruptly just short of the foul line. It does not have the heft and twist of a sixteen-pound ball torqued as it is laid down softly onto the lane and watched as it curves into the pocket to knock down all the pins at once. Or the sheer joy of defying gravity when scoring a soccer goal with a header or a scissors kick. The equipment may vary from culture to culture, but the universal sense of enjoyment that comes with playing with forces needs no translation.

It is a happy fact that students of force and motion can learn as much if not more from fiddling around with common objects and playing recreational sports as they can from just seeing them drawn as little arrows on a blackboard or hearing about them in words that issue from a teacher's mouth—or an author's pen. Participation in physical activity is a complement to the education of the mind, and it can provide a shortcut to understanding by subordinating abstraction to action.

Some people have moments of inertia that can last a lifetime, some seem to be just spinning their wheels, and others seem always to be moving and spinning and leaping like Olympic skaters. These are the movers and shakers who run for political office, invent new technologies, start up new businesses, play professional sports, and advance our arts and sciences. They work hard and play harder. As children they learned to enjoy dealing with the forces of the world, a skill that helps them balance a career and a life. Understanding the underlying role that force and motion play in so many of our life experiences helps us recognize that in being human a change of pace is not only welcome but also regenerative.

9
Forceful Illusions

Schools and Spools

Astounding performances from parlor tricks to high-wire acts exploit the mechanics of force. There is an old vaudeville act in which a performer maintains a large number of dinner plates spinning atop tall, slender, flexible rods. The phenomenon depends to a great extent upon gyroscopic forces that keep the wobbling plates from deviating too far too quickly from the horizontal; even more so, it depends upon the concept of conservation of momentum. As with many a magic, circus, or other stage performance, special equipment was often employed. The plates used typically had a dished out dimple or conical depression in their bottom to keep the stick centered under the plate.

The plates (sometimes bowls were used) could be put in motion by being placed on the tip of the stick and then tapped quickly but lightly along the edge with an open hand, the way a runway grazes the wheels of a landing airplane to get them rolling. Alternatively, the raised rim upon which a plate sits on a table was used to contain the stick, which could be moved in a whipping motion to get the plate spinning. Since plates and bowls have relatively large moments of inertia for their size, once the performer got them going they continued like flywheels to do so. By the time a good number of plates were spinning, the first ones naturally had slowed down and begun to wobble. This drove the plate spinner back to the first plates and with the hand tap-grazing the edges or whipping the sticks the spinner boosted things up to speed again—and started additional plates spinning. The act progressed to the point where a dizzying number of plates were spinning simultaneously, and since some were always beginning to wobble, the performer was running frantically around the stage—feigning being prompted by the gasps of the audience—to keep the plates from falling off the sticks, whose flexibility added another dimension of tension. The act often ended with the performer arresting each plate by grasping it as if it were a spinning Frisbee and ending up looking like a busser carrying an armful of clean plates to replenish a diminished stack under the counter in a busy diner.

There are many less frantic amusements. A familiar party challenge is to ask someone to balance two forks on the rim of a water glass by using only a toothpick. To the uninitiated, the task seems impossible; to the force-savvy, it is elementary. The trick is to intertwine the tines so that the fork handles form a V. Such an assemblage has its center of gravity on a line that bisects the angle formed by the forks. In other words, the center of gravity is located in the empty space between the handles. A toothpick stuck securely into the mesh of tines and aligned along the line of bisection will pass through the center of gravity, making it possible to balance the entire assembly on the rim of a glass. The forks will straddle but not touch the glass and the whole will appear to be defying gravity, when in fact it is exactly the force of gravity that makes the trick work. The effect can be enhanced by putting a lighted match to the end of the toothpick inside the rim and letting it burn down like a candlewick. When the flame reaches the rim, it will automatically extinguish and the magical assembly will appear to be cantilevered out from the glass. The only forces at play are the weight of the fork-and-toothpick assembly and the counteracting upward force exerted by the rim of the glass, both passing through the single contact point. The result is analogous to what Faraday illustrated with the loaded toy doll. It also explains how tightrope walkers balance themselves with a long pole.

When the mechanical principles involved are understood, the veil of illusion falls away from many an awe-inspiring performance. Analyzing it with quantitative reasoning also takes the mystery out of the mysterious—without necessarily taking the fun out of the function. One circus act that always fascinated me was the female performer who did spinning acrobatics suspended high above the ground only by her long, sparkling hair. I wondered, Why was it not pulled out by the roots? I also imagined how badly her scalp must hurt while she was swinging and spinning around up there. When I began to think more like an engineer, which means among other things thinking with numbers, I became satisfied that she was not hurting much if at all, and she was not going to go bald. Since the performer was usually a small, thin young woman, her weight was probably in the range of 100 to 110 pounds. Somewhere, perhaps in an installment of "Ripley's Believe It or Not," which I read religiously in the comics section of the Sunday newspaper, I learned that the number of hairs on a human head was estimated to range from 100,000 to 140,000, with the larger number incidentally belonging to blondes. So, if we take the average head of hair to contain about 120,000 strands and—for the sake of mathematical convenience and mechanical conservatism—take the weight of the woman to be 120 pounds, and if we assume that her weight is distributed equally among all the hairs on her head, then each strand of hair would have to support only about one one-thousandth of a pound, a force so small that it is difficult to sense by the average person. Putting a question in a quantitative context helps answer it unambiguously. Experiments on a single strand of hair have confirmed that it can actually support almost three ounces (three-sixteenths of a pound) of weight, which means that a full head of hair would support 22,500 pounds, which is the equivalent of almost two hundred 120-pound acrobats or two large elephants.

A somewhat related question has to do with a fakir reclining on a bed of nails and sometimes even letting a person stand on his chest. Again, putting numbers to the problem helps clarify and demystify what is actually happening. Let's say that the man is lanky, about six feet tall and weighing 150 pounds. Such a person might have a width across his torso of about 18 inches and across his legs a width ranging from about 3 inches at his ankles to about 5 at his thighs. Altogether, his body might have about 750 square inches of skin around which the forces of individual nails can be distributed. If spaced an inch apart, then there might be on the order of 750 nails on which he is lying. The force per nail is therefore on average about two-tenths of a pound. Concentrated on a nail point, which is more like a small platform, this is unlikely to produce anywhere near the pressure that it takes to pierce the skin, and so the fakir can lie upon the bed

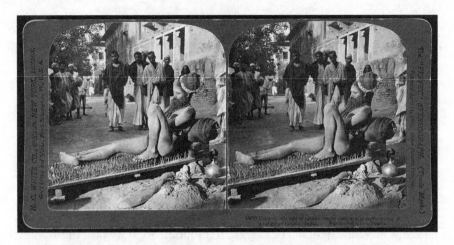

When viewed in a stereoscope, this pair of photographs, taken from slightly different angles and mounted on a card and known as a stereograph, will merge to provide a single three-dimensional image of the Indian fakir lying on a bed of nails. He is able to escape injury because the points of the nails are relatively blunt and the total force they exert is distributed across a relatively large area of his body, likely resulting in a force per nail measured in ounces rather than in pounds. *Library of Congress, Prints & Photographs Division, LC-USZ62-78659.*

of nails with barely a thought or sensation of pain. Of course, he would have to lie down on it with some care so that he does not concentrate too much of his weight on too few nails before assuming his final position. (If, for some reason, the fakir's bed had one nail higher than the others, he might feel it the way we can feel a single sprung spring in an easy chair and the fairytale princess felt one pea under several mattresses.)

Even if demystified, our appreciation of other circus and sideshow acts can be increased by a better understanding of the nature and magnitude of the forces involved and the skill needed to tame them. Many performances depend to a great extent on precise timing, especially when teams are involved in feats that involve motion. Consider, for example, the trapeze artist who swings toward her partner, who simultaneously is swinging on another trapeze toward her. The timing of the release of the first performer's grip on her trapeze and her grasp of the outstretched arms of her partner is critical. Neither the acrobat flying through the air nor the one preparing to make the catch tries to grasp just the other's hands or fingers, for that would amount to too small a target and leave little margin for error, not to mention relying on the weakest parts of an arm's anatomy. Rather, each performer reaches out for the forearms of the other and when contact is made lets the grip slide down toward the wrists. The long forearm not only provides a much larger and stronger target but also allows for a slight mismatch in aiming and timing. The transfer is typically done when the trapeze artists are at the high points in their swing, when like a pendulum at its apogee they are momentarily barely moving either up or down, and so providing an opportunity to make midcourse, split-second adjustments. If done quickly and with flair, the details of the move will hardly be noticed by the gasping audience. Understanding such tricks of the trade does not make the performance any less impressive. Indeed, recognition of the contingencies involved can make it even more so.

Just about every act in a three-ring circus involves the clever and careful use of and respect for forces and the motions associated with them. The trapeze artist sporting long arms, the tightrope walker using a balancing pole, or the strongman bending a length of steel—they all know when to be passive amid the forces at play and when to be active in producing the forces and movements needed to excite the crowd. Similarly, the unicyclist, the human cannonball, and the countless clowns crammed into a tiny car all know that the success of their act depends on how they deal with forces. So does the performer standing at one end of a teeterboard waiting to be propelled upward by the force of another performer jumping down onto the far end of the board, thus catapulting her onto the shoulders of the person atop a tower of acrobats. It is all about timing and taming, about understanding when to stand, when to walk, and when to run and jump, and letting the forces do their thing.

Feeling a force is not the same as having a feel for the force, especially in a context where distraction, deception, and duplicity are involved. A baseball catcher may feel the impact of a wicked fastball, but it is the pitcher who has to have a feel for the forces that cause it to do what he wants it to on its way to home plate. To pitch a baseball well involves having a feel for the ball in the hand. As the pitcher holds it behind his back waiting for a signal from the catcher, he rotates the cowhide in his hand until he has it in the proper orientation for the pitch. He feels the ball's stitches as they pass beneath his fingertips the way a blind person does the dots of braille.

If he is the kind of pitcher who enjoys having an extra advantage, he feels for the place on the ball that he nicked or roughened. Only when the ball feels correctly oriented in his hand to deliver a fast, slow, high, low, curve, sinker, slider, lob, spitball, or brushback pitch will he wind up and let go with a force that matches the choice.

To hit a baseball well involves the batter anticipating how the pitch will approach home plate. With so many options open to the pitcher, a guess is as good as a miss. The slugger will have eyes that can read the pitch in the half second—the blink of an eye—it takes the ball to travel the sixty feet from the pitcher's mound to home plate. During that time, he tenses the muscles in his hands, arms, shoulders, torso, and legs in a way appropriate to the situation. To field a well-hit ball, infielders and outfielders must also be able to make a quick calculation, and without the aid of a digital computer. Even mediocre players know to run to where the ball is going—or, in the famous advice of the hockey star Wayne Gretzky, know to skate to where the puck will be—but they do not always do so with enough precision. Those players with innate talent honed through practice will not only get to where the ball will be but will get there in just the right time to jump, crouch, or dive so that their glove is exactly where it needs to be to make the play.

Different sports naturally involve different kinds of balls, not all of which are spherical. In American football, the ball has the shape of an elongated sphere known as a prolate spheroid. Unlike the hard core beneath the stitching of a baseball, inside the laced-up leather skin of a football is an inflatable rubber bladder. Whereas a baseball pitcher grips the ball differently to deliver distinct pitches, a football quarterback generally throws all downfield passes with much the same grip. When the center hikes the ball to him, he makes sure his fingers are touching the laces; if they are not, he twirls the ball in his hands until they do. In this way the quarterback has his fingers in the optimal position to grip the football and give it a spiraling gyroscopic motion about its long axis, which keeps it from wobbling and makes it easier to catch. When a series of passing or rushing plays does not result in a first down, a punt or a field goal may be called for. If so, the ball is hiked to the punter or to a holder, each of whom attempts quickly to orient the ball with its laces away from the kicker's foot. If struck on the laces the ball could take an unpredictable turn.

A quarterback's pass is influenced not only by where his hand grasps the ball but also on how firm a grip he can get to impart the proper throwing force. Just as a young child can get a better grip on a football made of Non-Expanding Recreational Foam—a Nerf football—than one made of plastic, so a professional quarterback will be able to get a better grip on an underinflated than on an overinflated ball. To keep the game fair, National Football League rules require game balls to be pressurized to about thirteen pounds per square inch. In the 2015 American Football League championship game between the New England Patriots and the Indianapolis Colts, the Patriots were accused of using an underinflated ball, and quarterback Tom Brady was threatened with a four-game suspension. He fought the suspension for almost eighteen months before accepting the penalty.

In soccer, the ball is spherical, but it has no standout laces or stitches, which is just as well because, except for the goalie, players cannot use their hands to reorient the ball. This is not to say that players cannot choose what kind of force to impart to the ball. A glancing blow from the side of a shoe can make the ball curve around defenders and fly into the net. A successful shot depends not only on how large a propelling force is but also on how it is applied. This is true in sports ranging from bowling to billiards.

Outside of sports, there are activities in which a keen sense of feel is helpful in locating and orienting a specific object among many similarly shaped ones. A person standing before a vending machine should be able to extract from a pocketful of jumbled coins one of the needed denomination. There are many different currencies, of course, and they come in a variety of shapes and sizes. Although a circular disk is the most common, coins can also be polygonal and pierced, and their metal can be dense or light. A practiced hand can distinguish them by touch and heft. An American dime and quarter are easily distinguished by size and weight alone, and their ridged edge further distinguishes them from the smooth circumference of pennies and nickels. Thrusting a hand into a pocketful of change and pulling out a quarter properly oriented to be inserted into a coin slot is really no magic trick.

Engineers can also find things by touch and feel before they see the light of day. From experience, a structural engineer knows what arrangement of beams and columns will fill the gap in an existing urban landscape with a new skyscraper; a mechanical engineer knows what arrangement of links and gears will fit between an automobile engine and its driving wheels. For novel situations, engineers have to extricate from the jumble of experience they carry in their heads the odd idea that hitherto may have seemed like a foreign coin among those of the realm. They have to know how a mental image translates to a real structure or machine that will sit on the ground, ride on the roads, fly through the air, or soar into outer space. They can feel how it will behave when it is pitted up against the natural forces of gravity, wind, earthquake, and more. They develop a second sense that enables them to experience in their minds the force, stiffness, and motion associated with the thing before it is even made.

In early nineteenth-century America, before civilian schools of engineering were established, there were three principal ways to become an engineer. The first was to attend the U.S. Military Academy at West Point, New York, which offered courses in civilian engineering. The second was to pursue a course of self-study by reading about great engineering projects of the past and following the progress of ongoing projects. The third was to take a low-level job on a high-level project and, by demonstrating talent and responsibility, work one's way up the ladder. The construction of the Erie Canal provided such an opportunity, which is why it is referred to as the country's first school of engineering. It was only after midcentury that engineering colleges began to proliferate and provide a fast track to entering the profession.

A century later, after World War II had elevated the importance of high technology through such inventions as jet aircraft, radar, and nuclear power, there was great growth in the study of engineering. Returning servicemen and women, encouraged by the benefits of the G.I. Bill, swelled enrollments on college campuses and overloaded their physical facilities. To handle this influx, buildings for instruction and housing were quickly erected on ground cleared beside and beyond the tree-shaded walks of academic quadrangles. These structures, expected to serve only until the surge in enrollment subsided, often consisted of closely spaced Quonset huts and barracks. On some campuses, they remained in service for decades.

I took most of my engineering classes in a cluster of such buildings, which we unimaginatively called the barracks. It was impossible to ignore that we engineers (as we were called, even though we were really just students of engineering) were relegated to these shacks while our counterparts studying the arts and sciences did so in ivy-covered brick buildings of classical architecture retrofitted with the amenities of modern education. There was also the irony that we who were preparing to become members of a profession that was making technological leaps in commercial air travel, broadcast television, and satellite communications had to learn in the crudest of classrooms while the likes of English and philosophy majors were discussing in comfort the works of classical and medieval literature that espoused naive and long-superseded notions of force and motion.

But, after a week of freshman hazing, we engineers became so focused on the blackboard that we had little time to think about status symbols. Rather, we concentrated on learning the letters of the Greek alphabet, both upper- and lowercase, that professors of calculus, physics, and engineering used with abandon in the equations they chalked up on the board as if they were tallying the number of right and wrong answers to a trick question. The quality of their handwriting varied greatly. I recall one math professor introducing himself on the first day of class by saying his name and writing it out on the board. To me, it sounded as if he said the Greek letter xi (χ), but the chalk marks looked like "21A." After class, there was little agreement among us students over what exactly his name was. Consulting the course catalog we found that our instructor was a Mr. Zia. We young engineers had overlooked the obvious because we had already gotten so used to writing a Z as $\mathcal{Z}$, with the extra horizontal line distinguishing it from a 2. (For similar reasons, we learned to write zero as $\emptyset$, to distinguish it from the capital letter O. Likewise lowercase l was written as ℓ to avoid confusing it with the numeral 1.)

The writing of some professors was exemplary. I once observed surreptitiously a professor alone in an empty classroom standing before the blackboard and seeming to be writing line after line of something important, as if he were a misbehaving student sent to the board to write one hundred times, "I must not talk in class." On closer examination, I could see that he was writing letters of the Greek alphabet over and over. I learned later that he was a man who so prided himself in his blackboard penmanship that when he had some time before class he practiced his πs and 9s. His repetitive activity reminded me of learning to write in cursive according to the Palmer Method, which involved writing lines of interconnected loops and ovals by moving not just our fingers but our whole arm. We were taught writing as a physical activity, as something we could feel in our muscles, as a matter of force.

Before the widespread presence of white boards and laptop projection equipment in schools, you could count on a classroom to have at least one blackboard made of the kind of slate mined in Wales and written on with chalk like that from the White Cliffs of Dover. One talent we monolingual, urban, commuter students did bring to college was knowing how to use a piece of chalk, for we had quite a feel for the soft yet brittle material after drawing countless home plates, bases, and foul lines on city streets in preparation for a pickup game of stickball. We knew how to hold the chalk and how hard to press on it so it would not break. We could tell that teachers had no street smarts when they allowed a piece of chalk to break or make a sound more high pitched than the air raid siren tests of the fifties. We knew that the squeaking sound came from the alternating catching and slipping of chalk on the slate, because we had heard and felt it when we drew our first bases on the pavement. The trick to not sounding like a minor leaguer was to choke up on the length of chalk, hold it at about a forty-five-degree angle to the board, and pull and push with a force large enough to make a mark but not break the chalk. It was a skill we learned from experience but would not understand in theory until we had a course in the mechanics of force and motion. Mechanics, which the nineteenth-century polymath William Whewell called both the "science of force" and the "science of machines," became my favorite subject.

Among the things that engineers learn early on is that just about every technical field is analogous to every other technical field. And since mechanics is the oldest branch of physics, it can serve as a paradigm. Force—the word and the concept—kept cropping up in my courses in chemical, electrical, and even nuclear engineering. It was not only a piece of chalk that could break when subjected to too much force: Everything has a breaking point. But also, according to Whewell, "Every failure is a step to success," and this was another lesson I was to learn.

An engineer's college days are truly ones with a heavy workload, and those of us who thrived in such an environment tend to hold onto the ethic. In fact, some of us enjoyed school so much that we were loath to see it end and did all we could to extend it. Instead of taking a job upon graduation, we reupped for more schooling. We earned master's and doctoral degrees and, still not sated with the stuff of textbooks, went into teaching. Unfortunately, what many of us took into the classroom was little more than book learning. While we may have had a feel for the familiar, not all of us developed a very sophisticated sense of force and motion on an industrial scale. Unlike nineteenth-century engineers who applied their book learning to inventing new machines and building new infrastructure, many contemporary academic engineers apply their textbook learning to textbook writing. At first in my new position, I taught by the books. Only slowly was I able to wean myself and my students from them by drawing upon my limited engineering experience in holiday and summer jobs and some gained vicariously from reading about great projects.

However, after nine years of classroom study of engineering, eleven years of teaching it, and another five working on concepts never to be realized, I had to admit that I did not know in a broader sense exactly what engineering was and what engineers do that unites them and their profession, or at least I did not know this well enough to convey to a neighbor who was not an engineer how my profession fit into the larger scheme of things. Even though I was a registered professional engineer, I still felt like an apprentice. I manipulated equations of force and motion and distortion on paper and blackboard, but I was not confident that I understood their true meaning and significance in any broader context. As part of my college education, I had taken courses in the philosophy of science, which naturally led to my reading rather broadly into the history of science, but I still had trouble contemplating the nature of the universe and the forces it encompasses.

I tried to delve into the history and philosophy of engineering, but there was a dearth of material on those subjects in the library and at the bookstore. Those few books that I came across did not hold up against their counterparts in science. The easy conclusion was to consider engineering just a subset of science, but that position did not withstand scrutiny. The innovations of the Industrial Revolution did not follow from science alone. It may have provided inspiration for an idea and warnings of its limitations, but science soon had to take a back seat to some additional, practical forms of knowledge, intelligence, and process. Although the conventional wisdom seemed to be that science was in some way superior to engineering, evidence for that claim was also lacking. Yet scientists continued to consider it a given and asserted it through their words and actions.

It is not a new phenomenon. Among the books I had acquired in my quest for understanding was Heinrich Hertz's *Principles of Mechanics Presented in a New Form*, published posthumously in 1894. A preface by the physicist Hermann von Helmholtz, under whom Hertz had studied and served as a postdoctoral assistant at the University of Berlin, provided insight into the younger physicist's brief life and career: "At the end of his school course [at the University of Munich] he had to decide on his career, and chose that of an engineer. The modesty which in later years was such a characteristic feature of his nature, seems to have made him doubtful of his talent for theoretical science." Engineers do tend to be modest, allowing their works to speak for themselves; scientists, on the other hand, seem to have no "timid modesty" about their eponymous discoveries. Since Hertz was both engineer and scientist, it should come as no surprise that we know little of his time in industry but much of his scientific accomplishments, memorialized in the eponymous international unit for frequency: One cycle per second is known as a *hertz*, abbreviated *Hz*. Radio and television channels are broadcast within specific ranges of kilo- and megahertz (kHz and MHz).

The study of forces and how they affect motion was once considered part of natural philosophy, a term that began to be replaced by the word *science* in the seventeenth century—the time of Isaac Newton. Mechanics, the subdivision of the branch of science called physics, was used by the ancient Greek philosophers. Modern engineers are not known to be philosophers, but some of them do identify themselves as scientists and may do so in recognition of the close similarities in methodology that scientists and engineers follow in going from observation to hypothesis and from problem to solution. Regardless, there are two kinds of professorial engineers: those who think of and identify themselves as scientists (or engineering scientists) and those who prefer to call an engineer an engineer. These latter cringe when they overhear someone saying that an engineer is someone who operates a locomotive or supervises the maintenance and operation of a large factory, even though it can be a very satisfying and rewarding career to keep the train on the track and the building properly heated and cooled. Some engineers think it unfortunate that their work is associated with the term *mechanics,* if its use evokes images of automobile repair workers rather than professionals who study and understand forces and motions to such a degree that they exploit them to design powerful diesel engines, energy-efficient furnaces, and quiet air conditioners. Such engineers embrace mechanics as an efficacious analytical tool rather than as a pejorative.

The mechanical forces that interest engineers range from the familiar to the exotic; their effects can range from the insignificant to the explosive. An understanding of the mechanics of forces and their consequences is necessary for an engineer to design the structure of everything from a modest pedestrian bridge to a complicated space station, from a life-saving medical device to a life-ending weapon of mass destruction. A feel for forces is essential also for anyone wishing to understand the nature and limitations of the physical world of such things and their metaphysical implications. Without this, the world may end not with a whimper but in a bang.

History is full of lessons in the limits of human sense and sensibility when it comes to force. Among modern examples is the 1981 collapse of the elevated walkways of the Hyatt Regency Hotel in Kansas City, Missouri, a structural failure traced to a seemingly minor change in a connection detail that doubled the force it was designed to withstand. The technical issue was simple enough to assign as a homework problem to first- and second-year engineering students. Its solution relies not on calculus and advanced mechanics but on an understanding of the forces involved. Classroom narratives and homework problems such as this are captivating and provide lessons that every student should learn and no engineer forget.

Another example involves a pivotal moment in a meeting of the presidential commission investigating the 1986 explosion of the space shuttle

At a work site, a large spool of cable is typically mounted on a fixed axle so that the spool will not roll as the cable is played out from it. *iStock.com/RYosha.*

Challenger. Richard Feynman, professor of theoretical physics at the California Institute of Technology and a member of the commission, conducted a simple tabletop experiment during a public hearing. His equipment consisted of a cup of ice water and a C-clamp with which he compressed a small rubber gasket known as an O-ring, larger versions of which were critical to the shuttle's booster rocket design and the subject of some dispute as to their relevance to the failure. Whereas NASA managers claimed that the O-ring seals on the *Challenger*'s boosters were resilient and expanded to fill any gaps, Feynman demonstrated why that would not have been the case in the frigid conditions of the launch. When he took his O-ring out of the ice water in which it had been sitting and released the clamping force, he showed that the rubber remained deformed. In the actual rocket, this kind of behavior left gaps through which hot gasses escaped and led to the explosion. Simplicity clarified complexity.

Most professors know they are performers playing to an audience. During my acting days, I often used props to drive home the point that engineers

A spool sitting on the ground will be able to move as the cable is pulled off of it, but in which direction will it roll? *iStock.com/fermate.*

have not only to know but also to feel the forces involved in their designs. I have put on what might be termed little mechanical dramas, the sometimes surprising outcome of which can be eye-opening for students, engineers, and laypersons alike. It is my experience that even people who had taken several physics courses can have a surprising amount of difficulty predicting what will happen when a simple force is applied to a simple object in a situation just a bit more complicated than a vector in a vacuum.

In one illustrated lecture, I show images of large wooden spools of cable, the kind found around neighborhoods where internet service providers were upgrading to fiber optic and other advanced communication capabilities. The first slide shows several spools of black cable, one of which is mounted on a parked trailer. When the cable is pulled off the top of the core of this spool, it will rotate in place. If the cable were to be pulled in the same way from one of the spools shown sitting on the ground, the spool would roll. The second slide shows a spool of orange cable sitting on the grass beside a trench. There is a shovel in the background, suggesting that the trench had only recently been dug to receive the cable that is being fed off the bottom of the spool's core. I remind the class that engineers wishing to do a first-order analysis of the forces on and consequent motion of systems like these two spools will ignore irrelevant details, such as the color of the cable, the material of the spool, and to what purpose the cable is being put. What is relevant are the force pulling on the cable, the geometry of the spool, and the incline of the ground surface. The question I pose to the class is simply this: If the surface is level, the spool is circular, and the pull is parallel to the ground and away from the spool, in which direction will the spool roll?

After showing the images, I hold up a four-inch-diameter plastic spool that once held stereo speaker wire but now is wound with ordinary cotton cord. I tell the class I could more easily have brought a bobbin or a yoyo, but they might have been difficult for the shy students sitting in the back row to see. I ask all the students to imagine pulling the cord of my tabletop apparatus. Predicting which way the spool will roll when the cord is pulled is a question whose answer has eluded the unprepared minds of many an otherwise astute student of physics, fun, and games.

Before placing the spool down on the table before me, I poll the students: What will happen when I pull on the free end of the cord with a force parallel to the tabletop? Will the spool roll toward my hand or away from it? Or will it stay where it is, spinning like the wheel of a car on a patch of ice? And does the answer depend on whether I pull the cord off the top or off the bottom of the spool?

First, I ask for a show of hands about which way the spool will roll when the cord comes off the top. Virtually every student thinks that the spool will move in the same direction my hand does. After the vote, I demonstrate that they are correct: when I pull the cord, it unwinds from the spool, which in turn rolls toward my hand, though at a slower pace than my hand is moving away. Soon, my hand is so far from the spool that it is difficult to keep it moving in a straight line. We take a brief intermission as I retrieve the spool, rewind the cord onto it, and place it down on the same spot as it was at the beginning of Act 1.

In the case where the cord is pulled from the bottom of the spool, virtually all the students predict that the cord will also unwind from the spool, but it will roll away from my hand. Some of the students in the back of the room look puzzled and raise their hands tentatively. The final vote is split. Some students seem to suspect a trick question and vote against the masses; others think it could be a double trick and stick with their first guess. When I demonstrate that the cord will actually be wound up around the spool as it moves toward my hand faster than my hand is pulling on the cord, almost the entire class is visibly and audibly surprised—and skeptical.

Disbelieving what they see, some ask, "Where's the gimmick?" "What's the trick?" "Is the table imperceptibly tilted, making it an inclined plane, so the spool rolls downhill toward my hand?" I reply, "No," and to confirm my assertion I turn the spool around and pull the cord "up" the nonexistent slope. Even when we all see once again the cord being wound up onto the spool and it closing in on my hand, some remain unconvinced. I advise the whole class to try the experiment themselves in the quiet of their own room with any size spool and see if they can turn the tables. In the subsequent class, no one can report having gotten a different result with a different spool on a different table. As the curtain descends on Act 2 of the mini-mystery drama, I write on the board, "Newton's Second Law: An object moves in the direction of the force applied to it. $F = ma$." Force and acceleration are always in the same direction, regardless of the mass involved. Can you feel it?

I thought about my spool school when I came across an article in the *New Yorker* magazine about "paper-path engineers," those who tackle problems relating to jams in printers and copying machines. The article described a team discussing a problem encountered in a printing plant. Sheets of paper sailing from one conveyor belt to a higher one—in order to change which side will contact the next inked cylinder—were falling short of their target. A computer modeler who simulated the system, which employed a vacuum to suck a sheet of paper up and hold it onto the second conveyor, demonstrated that as the sheet passed from belt to belt its corners were drooping. It was as if a tired trapeze artist was not quite able to reach the outstretched arms of her partner.

After a wide-ranging discussion of how to keep the sheets flat, the lead engineer exclaimed, "Bernoulli!" His idea was to shoot a jet of air over the drooping corners, so that they would be lifted. Voila! The writer of the article explained correctly that Bernoulli's principle relates air speed and pressure. In a manner analogous to the bank effect that contributed to the container ship *Ever Given* going aground in the Suez Canal, the faster-moving air would create a lower pressure on the top of the sheet of paper than exists on its bottom, a situation resulting in a force that aeronautical engineers call lift. Unfortunately, the writer took his explanation of Bernoulli's principle too far: "Because the top side of an airplane wing is flat, while the underside is curved, the air above moves faster than the air below, and the wing rises."

I wondered, was this the kind of simple mistake we all make when typing a sentence? But if it was, how did it get past his proofreading of his manuscript, past an editor's pencil, past the magazine's fact checker, and its own proofreader? Were they all paying such close attention to spelling and grammar that they did not see the sentence for the words? Perhaps none of them ever took a physics course, but knowledge about the world does not come only from formal education. An understanding of a physical phenomenon comes also from observation. So, I wondered further if the writer ever sat in a window seat and looked out at the wing of the airplane. He very well may have, but what did he see? He should have seen the wing, of course, but he could not have observed it very closely or, if he did, remember correctly what he saw. Surely the real wing did not have a flat top. If it did, it would present a shortcut for the air traveling from front to back, meaning that the air split into two streams would flow faster across the bottom than over the top before coming together again at the trailing edge. According to the principle articulated by Daniel Bernoulli, who applied mathematics to the mechanics of fluids in his 1738 treatise *Hydrodynamica*, speed and pressure are related in such a way that the slower a fluid moves across a surface, the higher the pressure it exerts. A flat-topped wing would thus be pressed downward! If such a plane ever could get up off the ground, it would be immediately pushed right back down onto it. The Wright brothers, neither of whom completed high school, came to understood this, and by making the shape of the planar wings of their Flyer adjustable, they achieved controlled flight by warping them into a proper configuration.

The *New Yorker* writer's mistake did not escape Jim Stoffer of Astoria, Oregon, a reader who remembered from flight school that it was the top and not the bottom of an airplane wing that is curved. And he noted that it was the Bernoulli effect that also enables sailboats to "tack into the wind."

Numeracy, the ability to think in terms of quantities, including geometric measures, and their comparison, is the scientific counterpart to literacy. Looking at something and not seeing it as it is can be just as much a hindrance to understanding the ways of the physical world as not having a vocabulary sufficient to grasp the meaning of words in a novel is to getting the full story. Even graphic artists, whose stock in trade should be having an eye for how things really look, make surprisingly many mistakes in depicting something as common as a sharpened wooden pencil. I have documented this elsewhere, so let it suffice to say how ironic it is that many incorrectly rendered pencils are actually drawn with a pencil, with a physical model likely closer at hand than an airplane wing is to a window seat. Of course, if artists are so confident in their preconceived notions of what a sharpened pencil looks like, they may very well just see the real one in their hand that way, too, the way the *New Yorker* writer saw the airplane wing as flat-topped.

It used to be the case that among the first courses engineering students took was one called mechanical drawing, in which images of structures and machines were executed employing T-squares, right triangles, and French curves. These mechanical devices—as well as the subject matter of many an engineering drawing—may have given the course its name, but it is a misnomer. Before the advent of the digital computer, at least, the drawings students were challenged to produce were not done by a machine but by the hands of the engineer. And the engineer was taught to make the drawing match the subject exactly, so that if necessary the artifact could be manufactured properly from the drawing alone. If a concept for a device or structure is not rendered realistically, it might not be realizable at all in a machine shop or on a construction site. Today, drawings are done digitally, which is more mechanical than manual, and those operating the digital machine may never have really looked at the real thing or calculated the implications of the drawing. If there was a prototype sitting beside them, they might never even have touched it—or wanted to.

10

From Physics to the Physical

The Real Feel

In 1773, the French polymath Pierre-Simon Laplace asserted that if the mass and velocity of every object, as well as the forces acting on it, were known at a single moment, then the future configuration of the universe would be perfectly predictable. As he put it in his *Philosophical Essay on Probabilities,* "We may regard the present state of the universe as the effect of its past and the cause of its future. An intellect which at a certain moment would know all forces that set nature in motion, and all positions of all items of which nature is composed, if this intellect were also vast enough to submit these data to analysis, it would embrace in a single formula the movements of the greatest bodies of the universe and those of the tiniest atom; for such an intellect nothing would be uncertain and the future just like the past would be present before its eyes."

The all-knowing intellect came to be known as Laplace's demon. Today, we might refer to it as a super-super-supercomputer, one capable of accepting and processing such an unprecedentedly large volume of data that just writing out the number in the conventional way would fill many books. To simplify the writing of very large numbers, engineers, scientists, and mathematicians use so-called exponential notation, in which a number such as 1,000,000 becomes 10^6, that is, ten raised to the sixth power, or a one followed by six zeroes. Many large round numbers are given names: 10^6 is called a million, and 10^9 is a billion. (Until 1974, in the United Kingdom a billion meant a million million, or 10^{12}, which made the term ambiguous to Americans.) The number 10^{100}, which can be written out with a hundred zeroes but hardly comprehended, was known technically as ten duotrigintillions before it was called a googol, a word coined in 1920 by Milton Sirotta, a nine-year-old nephew of the American mathematician Edward Kasner. A mishearing and consequent misspelling of this word resulted in the name Google for the now-dominant search engine whose mission is "to organize the immense, seemingly infinite amount of information available on the Web."

But whereas words are cheap, even ordinary supercomputers are not. Rather than compete with Laplace's demon, less ambitious scientists and practical engineers have looked to universes comprising a smaller number of objects, such as the triad of the Earth, Moon, and Sun, to demonstrate the complexity of the full one. But finding even an approximate solution to the so-called three-body problem is computationally intensive. Only a two-body problem, such as the one involving the linked motions of the Earth and Moon, can be solved in closed form—that is, as an exact mathematical equation rather than as a highly accurate but inexact digital output. A problem involving a single force and its effect on a single object falls into the exactly solvable category.

Ardent advocates of large computer models believe that, given sufficient storage and memory, they can, like Laplace's demon, solve virtually any problem. Experienced engineers are wary of such claims. The distinguished structural engineers William LeMessurier and Leslie Robertson on separate occasions told me that they never accept outright the results of a computer simulation brought to them by a young associate; they make a hand calculation based on a simple model. For example, it is important to know at an early stage in a tall building's design how much a proposed structural concept will allow the building to sway in the wind. The experienced engineer with a feel for forces and their effects can estimate the amount with a pencil-and-paper calculation involving a system consisting of a single force acting horizontally on a vertical cantilever anchored in the ground. If results of the pencil-and-paper calculation and the computer simulation are close to each other, the veteran engineer gains confidence in the black-box computer model; if they are not, the engineer-in-training will be sent back to the digital drawing board.

Once, when I was teaching a graduate course in elasticity, a subject that relates the forces acting on an object to its shape and stiffness, I assigned as a homework problem the determination of the so-called stress concentration factor for a circular hole in a strip of material that is being pulled, the way a belt would be stretched by someone who had gained weight since last trying it on. The exact solution was readily obtained mathematically with the methods taught in the course, and that is how most students approached it. However, one student, who had come back to school for a master's degree after working a few years in the aerospace industry, used a computer model. When I congratulated the students who got the exact answer, which is 3, the student who instinctively went to the computer challenged us. His digital calculation had resulted in an answer of 2.541, which he thought was more accurate. I suggested that he might refine the representation of the pierced strip he had input into the computer model. At the next class he proudly announced that the more refined— the closer to reality—he made his representation, the greater was the multiplier produced by the computer. In fact, he confessed that the number looked as if it was approaching 3.

Regardless of the number, the result explains why an old leather belt pulled too tightly will break across a hole and why perforations provide a preferred path along which a sheet of paper will tear. It also explains why an overpressurized airplane fuselage will rip open along a line of rivet holes. Since the intensity factor at the tip of a sharp crack is even greater than 3, blunting the crack by drilling a small hole at its tip can stop it from growing in length across a bicycle fender, a trick known to seasoned cyclists and aircraft maintenance workers alike.

The way engineers think about forces and their effects necessarily affects how they think about things in general and the interaction of specific things in particular, including the interaction of people with things. To the engineer, forces animate the world and explain its workings, at least on a physical level. Through the language of idiom, meme, and metaphor, they can also influence how we think about the mental and metaphysical. Oliver Wendell Holmes Jr., an associate justice of the U.S. Supreme Court, once wrote that the "great forces that are behind every detail" make "all the difference between philosophy and gossip." In the world of engineered structures, the forces do not have to be outsized to influence the way a structure behaves. A small crack in a small hole gone undetected can bring down an airplane. The individual parts for a new airplane design may be tested individually and prove to be strong enough for their specific role in the structure, but that is not considered sufficient to prove the integrity and operability of a fully assembled plane, which is why it must be flight-tested.

Interestingly, the jurist Holmes's reflections on forces and details were anticipated somewhat by his own father, the nineteenth-century physician and poet Oliver Wendell Holmes Sr. His poem "The Deacon's Master-piece," also known as the "One-Hoss Shay," recounts in verse the story of a carriage commissioned by the deacon. Understanding that "a chaise *breaks down,* but doesn't *wear out,*" because "there is always, somewhere, a weakest spot," the deacon had a wagon built in which each part was equally strong. It was built "in such a logical way" that "it ran for a hundred years to a day" and then suddenly collapsed in a heap.

No matter how complex a system an engineer is called upon to design or analyze, it can be modeled and each part evaluated in terms of the forces imposed upon it by adjacent parts and the consequent forces it exerts on other adjacent parts. By identifying, understanding, and quantifying the forces that act between beams and columns or wheels and axles, an engineer can assess whether a particular selection and arrangement of them into a specific structure or machine will be strong and safe enough for people to use. Knowing the exact value of each force would be nice, but since a multiple of it will be employed to ensure safety, a good approximation can do. Safety factors vary greatly with context. For an elevator cable, it may be 5 or more; for an airplane, a factor of more than 1.5 might mean that it would be too costly to build or too heavy to fly.

Such details are extremely important to engineers, who always prefer to err on the side of safety. Experience has taught lay users of engineered structures and products that they are overwhelmingly safe. Indeed, engineers adhere to a code of ethics that requires them to "hold paramount the safety, health, and welfare of the public," which is akin to medical doctors taking the Hippocratic Oath to "first, do no harm." People are accustomed to trusting professionals with their lives.

Should consumers wish to understand why a piece of furniture does not trap them or collapse when they sit and recline in it, they can and do look into its design the way many medical patients look into the details of a drug, treatment, or procedure suggested by their doctor. Having a feel for the way the individual parts of a mechanism react to the forces and motions to which it is subjected should be reassuring. The working parts of a recliner, for example, are typically elemental ones such as steel bars, bolts, rivets, and springs. The forces acting on any one of them come from the pull of gravity, the push of the person reclining, and their connection with other parts of the mechanism. In fact, regardless of the object, it is the forces that manifest themselves when it is in contact with another object, including the person using it, that govern its ability to perform its function as part of the whole, and do it safely and reliably. And this holds for everything we touch, whether we think about it or not.

We have become so accustomed to grasping and twisting doorknobs, pushing and pulling shopping carts, packing and unpacking grocery bags, pinching and swiping credit cards that we seldom reflect upon or even recognize that all these actions are possible and successful because we are capable of applying just the right kind of force in just the right amount to accomplish the task at hand. Specific arrangements of force are essential whenever people engage in such diverse tasks as operating specialized gadgets in the kitchen, handling classic eating utensils at the dinner table, playing an instrument on a concert stage, touching the display of a digital device, swinging a five iron on the golf course, closing a deal with a handshake, doing virtually anything with anything. And we do learn to apply forces with considerable appropriateness to the task at hand. Too light a grip on a carton we are carrying can result in it being dropped and bouncing down the stairs; too casual a hold on a hammer can allow it to drop on our foot or fly through the air and strike someone behind us. Conversely, pounding with a tight fist too vigorously on a window can break the glass, resulting in lacerations; using the wrong drill bit can result in it snapping off and striking someone in the eye. Everything has a proper use and a limit to how aggressively it can be used—a breaking point—including our own fingers, toes, arms, legs, skull, neck, and back, their bones being able to take only so much force.

Many a decision in a product-liability lawsuit boils down to whether the design or the user was at fault in causing an injury. Did the engineer properly anticipate and account for the forces under which the product was intended to be used? Did the consumer use the product so that the forces applied to it were within stated or reasonable limits? Engineers are often called as expert witnesses in such cases, but it is the lawyers and judge who control the narrative.

In the course of living, I have stubbed many a toe and broken some. When the stubbing involved a small force, it resulted in little more than a fleeting pain. When the force was large enough, it caused a fracture, and sometimes the pain distracted me from recognizing until the bone had healed that the toe was left deformed. In walking about for weeks on a broken toe, I had allowed forces exerted by my shoe to reshape the digit. The mathematical biologist D'Arcy Thompson, in his erudite treatise *On Growth and Form,* demonstrated how considerations of physical force alone can explain the shape of living things. Trees growing on a windswept coast develop trunks inclined away from the sea and bear asymmetrical branches and foliage. In the introduction to his book, Thompson sets out his thinking about the physical concept by asserting that "force is recognized by its action in producing or in changing motion, or in preventing change of motion or in maintaining rest," which is essentially an acknowledgment of the relevance of Newton's first law to living things. Thompson goes on to say, "Force, unlike matter, has no independent objective existence." In other words, we come to know force not by what it is but by what it does. This is certainly how engineers learn to think about the elusive thing. Furthermore, Thompson alerts the reader that when he uses the term *force* in his book, he uses it in the prenuclear sense, "as the physicist always does, for the sake of brevity." Engineers do the same thing, and it is in Thompson's sense that the term and concept *force* has been used in this book.

We may not think about forces per se, but their effects seldom cease to fascinate us. Infants will stare and smile at a colorful mobile hung above their crib, watching its lazy and comforting motion. I am no psychologist, let alone a child psychologist, but I know enjoyment when I see it. Their eyes are bright with interest and alert with attention. Something about the movement of the mobile's parts is at once enticing, satisfying, and amusing—and confusing. The slightest movement of the air, even when caused by the exhalation of a baby's breath or the wave of a tiny hand, can excite the mobile and initiate motion or cause a change in it. This is force at play, and infants soon catch on to the connection between their motion and that of the toy they reach for only to push it farther away. But the joy of movement can be as pleasurable as the thought of possession.

Older children do not appear to think a great deal about how they learn to move their hands and fingers in order to grasp a bat or ball or how to make the bat hit the ball or how to throw the ball so that the bat misses it, but they all do learn these skills to varying degrees. Whether those who get good at playing ball think more consciously about how they achieve what they do may be an arguable point, but many of us have experienced the deleterious effect of thinking too much about what comes naturally— or unnaturally. Not only can it become more difficult to throw, hit, or catch a ball when we overanalyze the mechanics of the action, but also we can diminish the enjoyment of the act itself. But observing and analyzing how someone else does something can help us understand why it is not as mysterious or as painful as it might look. Still, it is one thing to walk and chew gum at the same time, another to explain the mechanics of those movements.

An unsophisticated, seemingly common-sense understanding of force and motion is characterized by psychologists as a manifestation of "naive" or "folk" physics. In some cases, the ideas are remarkably similar to those held by Aristotle and other ancient philosophers who struggled with the nature of physical phenomena and believed, for example, that an object moving through the air continued to do so only because of the continued presence of some kind of pushing force. Even modern university students who have taken a physics course or two can reveal a tinge of such thinking. That the idea may be neither easy nor obvious is evidenced by the fact that the evolution of concepts of force has spanned millennia and is considered by some physicists today still to be evolving. It took the likes of Galileo and Newton to articulate the concepts of force, motion, and inertia as we understand them today.

As ontogeny recapitulates phylogeny in biology, an individual child's idea of force can grow the way civilization's has. My concepts of force and motion—had I been asked to express them as a young child—would certainly have fallen into the folk physics category. My playmates and I just did not give much thought to such questions as why a ball after leaving my hand followed the trajectory that it did. It did what it did, and we were perfectly happy to master the acts of throwing, catching, and hitting the ball with a bat, even if the ball was just a common Spalding rubber one and the bat was a sawed-off broom handle. It is remarkable what can be achieved with little or no sophisticated equipment or knowledge of the forces or motions involved in an activity as common as playing baseball or stickball.

Consider how quickly older children learn to throw a ball accurately to a playmate standing dozens of feet away. By means of a push of the hand they impart to the ball an appropriate speed and inclination so that it ends up close enough to the playmate that not a single step has to be taken to catch it. The act of catching an arcing ball is something equally admirable, for it requires the hand or glove to be positioned very near to where the ball will be at the end of its descent and to exert a sufficiently strong push against the impacting ball to stop it. Likewise, hitting a moving ball by swinging a bat is a remarkable feat of eye-hand-muscle coordination. Achieving accuracy in such tasks predated the games of catch and baseball, for similar forces and motions had to be mastered by the young of our prehistoric ancestors to throw a rock at a scampering rabbit, to catch a bird in flight, or to spear a fish swimming in a stream.

In spite of my having had an invisible friend in Gravity, my own naive perceptions of force and motion must have prevailed until I took physics in high school and college. It was there that I learned what the formal language of science called these concepts and how they were manipulated in the mind and on paper to explain phenomena and calculate the quantities involved. There was little if any delving into the philosophical meanings of force and motion, which were taken more or less as givens—as primitives, a philosopher might say. We learned to visualize forces and motions as directed lines (arrows known as vectors) that have mathematical qualities that enabled us to add them together and compare and relate them to each other and to their effects. The content of physics courses was not naive, but neither was it especially concrete. It was mathematical and abstract. Arrow forces attacking mathematical points produced arrow motions.

Forces and motions took on more concrete and realistic meanings in my engineering courses. A force was no longer just an arrow pushing a point around. As engineering students we learned that forces had consequences beyond motion. Instead of the perfectly spherical balls idealized as dots that illustrated the pages of our physics textbooks, our engineering texts depicted objects as irregularly shaped as potatoes that could represent anything from an errant asteroid to a precision machine part. The unbounded universe of physics had expanded into the constrained world of engineering. And a new language went along with the new geometry. Objects were no longer idealized as spheres and strings; now they were balls and chains. We learned that how these parts were connected to each other determined how they exerted forces on each other. Forces became the engineering abstractions of tangible things such as beams and columns and screws and bolts and nuts and washers and springs and cotter pins, not to mention electric currents and heads of steam. We learned how to disassemble structures and machines on paper and lay out each individual part in a neat drawing that represents abutting parts through the forces they imposed on it. Such a drawing constitutes what engineers call a "free body diagram," with the word *free* implying that although the part is shown separated from the rest of the universe, it remains connected to it analytically through the forces (represented by arrows) that they exert on it.

In many an engineering problem, the object is to calculate the magnitude of the connecting forces. Knowing their values means knowing how strong a part has to be in the actual structure or machine. To an engineer, forces are not just abstractions but are also the heart of the matter. Once the forces are known, the necessary size of a beam or bolt or spring or pin can be determined, albeit sometimes in a roundabout manner. What makes the problem tricky is that the calculated forces can be too large for a part to handle. Substituting a stronger part typically means using a heavier one, which alters the connecting forces, which then must be recalculated. These recalculated forces affect those on other parts in the system. A seemingly straightforward design calculation thus becomes one of iteration, a tedious process more suited to a digital than a human computer.

Unlike in physics, where the main problem is to explain the universe that is given already assembled, in engineering the main problem is to come up with something that exists nowhere in the universe until it is designed and manufactured, thereby mating force with function.

11

Forces on Inclined Planes

Unlevel Playing Fields

The sides of the Great Pyramid at Giza are inclined at about a 52° angle off the horizontal. This translates to a rise-to-run ratio of about fourteen to eleven, which is roughly twice as steep as that allowed by international building codes for a stairway in a home. We know how tiring it can be to climb up several flights several times a day; moving a piano up that same distance is understandably exhausting; moving two-ton blocks of stone up the sides of a five-hundred-foot-tall pyramid would seem to be virtually impossible. This, among other factors, is why there is general agreement among engineers that the Egyptian builders must have used ramps in some way. A pyramid may be thought of as a quartet of narrowing ramps meeting to form an apex, but given their steepness, it seemed unlikely that they could have been used directly as inclined planes up which heavy blocks of stone could have been pushed or pulled, at least easily.

Engineers consider an inclined plane to be a simple machine that somewhat tames the force of gravity. It takes many forms, including the playground slide and the wheelchair access ramp. An object placed upon an inclined plane will remain there as long as the bottom of the object and the surface of the plane are sufficiently rough and the slope is not too great. The phenomenon can be demonstrated by placing a brick on a board and slowly raising one end. The brick will remain in place until the board makes an angle with the floor at which the friction force between brick and plane no longer equals or exceeds the pull of gravity down the plane.

Engineers idealize the geometry of an inclined plane as a right triangle,

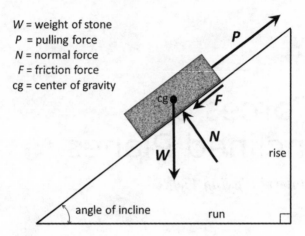

An engineer's abstraction of a block of stone being
hauled up an inclined plane includes a depiction of all
the forces acting on it. *PowerPoint drawing.*

and they see an object hauled up it as subject to three kinds of forces:
weight, pull, and reaction, this last representing how the surface pushes
up against the object and against its being moved. For the convenience of
calculation, engineers break this reaction force into two component parts,
one normal (perpendicular) to the plane and one (the friction force) parallel to it. As we have seen, the normal and friction forces are related through
a coefficient dependent on the nature of the contact surfaces. The mechanical advantage of the plane is that only a fraction of its weight has to
be overcome in moving a block of stone up the slope. The friction force
opposing movement must still be overcome, of course, but in reality that
can be reduced by proper lubrication of the surface.

Pushing or pulling a weight up an inclined plane may be no less stren-
uous than carrying it up a flight of stairs reaching the same level. In each
case, the person doing the pushing, pulling, or carrying must provide the
same energy to raise the weight to the same height. Eliminating even that
effort is the kind of challenge that inventors welcome. Wouldn't it be nice
if a flight of stairs itself moved, so all that a person had to do was stand in
place to be elevated?

The first U.S. patent for the now-familiar escalator appears to be one
for "revolving stairs" issued in 1859 to inventor Nathan Ames, of Saugus,

Massachusetts. As is common with inventions, his was followed by successive improvements made and patented by other inventors, but most of theirs, like his, were not built in their time. An exception was the "endless conveyer or elevator" invented by New Yorker Jesse W. Reno and installed at Coney Island in 1896. However, it was not Reno or any other individual inventor who coined the word *escalator*. That was done by the Otis Elevator Company, presumably to distinguish the inclined system of movement from the vertical one. At first, the U.S. Patent Office did not allow the neologism as a descriptive term, but it was registered as a trademark the same year the company's first device by that name was put into public use at the 1900 Paris Exposition. Today, perhaps the most conspicuous moving stairway in the world operates in Paris, in the form of the escalators cascading within transparent tubes cantilevered from the facade of the Centre Georges Pompidou.

A modern escalator is the setting for Nicholson Baker's short novel *The Mezzanine,* in which the protagonist is a young office worker named Howie. After lunch each day, Howie rides the moving stairs up one level to reach his office. On the day the story takes place, he reflects upon some mundane things other escalator riders might not notice. He finds it remarkable how a maintenance man standing at the base of the stairs is polishing the entire handrail by just clasping his cleaning rag around it as it moves like a conveyor belt carrying smudges and germs. Had he been cleaning the stationary balustrade around the mezzanine, he would have had to walk all along it to accomplish essentially the same end and feel the same forces. That is the nature of relative motion and the consequence of the ever-present pairing of action and reaction.

Typical escalator riders, especially if they are using a cell phone, are as oblivious to the details of the machine and its workings as they are of the surroundings through which it passes. Veteran riders may not even recognize that the only difference in sensation between ascending and descending on an escalator is the act of getting onto and off of it. It may never occur to them that in going up they are being transported against the force of gravity and in going down moving in concert with it. In between landings, riders on a moving stairway might as well be standing still on solid ground. They may ignore the abrupt changes they feel in boarding and leaving the escalator, but their body will feel the accompanying forces and unless it is unconsciously prepared for them, they may lose their balance momentarily.

A "moving pavement" was installed at the same Paris exposition that boasted the first escalator. The elaborate installation involved three parallel portions of sidewalk, the first of which moved at the normal speed of a pedestrian on solid ground; the second was fixed, like a train platform, to facilitate transfer to the third, which traveled at about twice the normal rate of walking. Unlike today's familiar moving walkways, which we stride onto in the direction of motion, those in Paris were mounted from the side.

Mounting or dismounting such a contraption was not considered very difficult for people accustomed to jumping on or off a moving streetcar or cable car or merry-go-round. None of those experiences was very familiar to me when I visited the gigantic Ferris wheel known as the London Eye, but it demanded the same skills of its riders. Except for stopping to board or unboard a disabled rider, it turns continuously; the exiting and entering of the moving cars at speed increases the number of people who can be accommodated on the tourist attraction.

Ever since childhood I have remained somewhat tentative in my approach to a moving staircase. In my first encounter with one, I felt that a rug was being pulled out from under me as I stepped from a solid terrazzo floor onto a segmented moving one. To the annoyance of seasoned shoppers and travelers behind me, I am sure, I tend to break step and pause a bit before boarding and hesitate upon leaving any moving device. I am equally tentative with how I tow my rolling luggage onto and off the step behind me. But thinking too much about the forces of transition can be as dangerous to travelers as mulling too much over the mechanics of their actions can be to athletes. It is one thing to reflect upon the forces involved in everyday activities, but thinking too analytically about them as we are being subjected to them can throw off our game and lead to disaster.

The problem of getting onto and off of any one of these people movers is generally more easily accomplished after negotiating an escalator, but when coming off a long airline flight and still seeking our land legs, the challenge can be noticeable. I have found it especially pronounced when people who have been traveling for hours in a pressurized cabin at over five hundred miles per hour cannot get their bodies or minds to adjust to a normal walking pace. They want to move as fast as possible to the airport parking lot and quickly drive home.

Some of the moving walkways in my home airport are not entirely flat, because ramps ease the transition for walkers propelling themselves along the length of the intermittently sloping concourse. People walking on solid ground negotiate these ramps as easily as they do a gently sloped driveway in summer, when it is not likely to be iced over. However, the traveler standing and daydreaming on an airport's moving walkway can be taken by surprise by the transition between the horizontal and inclined. On one occasion, when I had left my luggage standing alone behind me, it fell over when we reached the beginning of an incline. Curiously, whereas an audible warning alerts a person of the approaching end of a flat walkway, there is no aural warning for these changes in slope. I have seen other travelers get thrown off balance, like me, when experiencing the simultaneous effects of inertia and a changing support force.

Jet bridges are essentially inclined planes. Their angle has to be adjustable to accommodate aircraft ranging in size from small commuter jet to jumbo jet, which means that the slope of the bridge can range from steep to flat. Even the frequent traveler may find this disconcerting in presenting another challenging transition. The unsupported length of some jet bridges can be quite great, which allows a degree of bounce. The windowless bridge may also be thought of as a telescoping tunnel, through which the boarding passengers walk down like cattle into a crowded corral. The short transition pieces required between telescoping sections can be especially tricky, these short connecting ramps being of a noticeably greater slope. Passengers pulling roll-ons down the bridge may feel that their luggage wants to overtake them, pushing them downhill like a cart behind a horse. On the return leg of their trip, it may feel as if the bag is holding them back.

Once passengers are seated on an airplane, they may be lulled to sleep by so gentle and smooth a journey from gate to tarmac that they are not reminded of the surprising and exaggerated changes in force and motion they will feel during the flight. Their bliss may be prolonged if the plane taxis uneventfully to the runway, but once in place for takeoff anything goes. Some pilots keep the brakes on as the engines are screaming externally and the passengers internally to release the brakes and let the plane leave the airport. When the pilot does so, the plane leaps forward like a pit bull unleashed. As the plane accelerates down the runway, the passengers get a complementary jolt of inertia that pushes them back into their seat the way a smoking fast ball burrows into a catcher's mitt. The passenger who had been sitting comfortably but motionless cannot help but feel the full force of Newton's law. Whether the passengers feel this as the seat back pushing them forward to keep them up with everything else in the cabin or they feel they are pushing back on the seat, they are experiencing a force that seems to be saying, "Wish you were here." Luggage in the overhead bin that has no assigned seat slides backward, and improperly secured tray tables flop down.

On a Southwest Airlines flight once, I saw a demonstration of the combined effect of the force of inertia and the plane's angle of ascent when a flight attendant sitting in a jump seat near the forward galley strewed bags of peanuts on the floor. She challenged passengers in aisle seats to grab a bag as it slid past, but few of us were able to catch anything traveling so quickly down the inclined plane. I was unsure of exactly how steep our climbing angle was until I noticed a luggage strap hanging down from an overhead bin. During the stop and go of taxiing it behaved like a pendulum, but after takeoff it served as an inclinometer. The strap, which hung perfectly vertically when we were sitting at the gate, had departed increasingly from that position as the plane accelerated down the runway and climbed. The strap's angle relative to the reference edge of a window was determined by the position at which the forces of gravity, inertia, and tension in the strap were in balance. If it were raining, the water streaking across the window would do so at the same angle. Once the plane leveled off to cruising altitude and was no longer accelerating, the strap returned to the vertical, its tension force alone now being sufficient to counter the gravitational.

When an aircraft is flying through thick cloud cover or other low visibility conditions, the orientation of the entire plane can be difficult even for the crew to judge. Experienced pilots can become so spatially disoriented that they lose a sense of altitude and of up and down. This phenomenon is believed to have been the cause of some infamous aircraft accidents. In 1999, John Kennedy Jr. was using visual flight rules in piloting a single-engine plane from New Jersey to Martha's Vineyard off the coast of Massachusetts. When darkness and bad weather had combined to obscure familiar landmarks, the plane crashed into the ocean. The impact may have been the only unusual force he, his wife, and his sister-in-law experienced. Spatial disorientation was also blamed for the 2021 helicopter crash that took the life of basketball star Kobe Bryant, his daughter, and seven others, including the pilot.

Typically, an airline pilot will announce the start of a descent, and the subsequent sound of landing gear being deployed signals that the craft will soon be touching down. The descent can seem to take an excruciatingly long time, especially in weather where the ceiling is measured in hundreds of feet. Passengers looking out a window may see nothing but whiteness until the ground suddenly comes up into view and shocks them with its proximity. Passengers not looking out a window are greeted with a surprise jolt, for putting a plane down on a runway brings all sorts of new forces into play. When the landing gear is fully deployed, its wheels are not rotating. Since the runway is not moving, there will be a mismatch between the translating wheel and the stationary tarmac, which has to exert forces on the tires to get them up to rolling speed. Passengers experience this sequence of forces as, first, a bump as the tires hit the ground and, second, the screeching sounds of rubber as kinetic friction increases the rate of rotation of the tires until it matches plane speed, at which time static friction insures smooth rolling to the gate.

The concentration of black skid marks at the end of a runway attests to the fact that rubber is consumed in the process. The effect can be lessened by getting the tires up to speed beforehand, which can be done by fitting landing gear with motors to do so. Alternately, rubber flaps can be incorporated into the sides of the tires, so that they will act like sails in catching the wind and thus begin to spin the tire before friction from the runway surface does. In the meantime, the plane's wing flaps are deployed to present a different kind of sail to the wind, one that serves to push the plane in the opposite direction of its motion to slow it down. Sometimes the engines are reversed until the plane is moving slowly enough to switch them back to the forward thrust that carries the plane across taxiways to the gate. Throughout all this, the force of inertia will again have come into play, having thrown passengers forward against their seat belts and sending luggage toward the front of the overhead bin. These are some of the same forces passengers felt and will feel in negotiating the inclined plane of the jet bridge and on the moving walkways, escalators, and elevators that they will once again board.

12

Stretching and Squeezing

Springs and Packaging

In 1971, the astronaut Alan Shepard carried some golf balls and the head of a six iron on the Apollo 14 mission under his command, and while on the Moon he used the handle of a lunar-sample scooper to improvise a club. According to his own report, one of the balls he hit flew for "miles and miles and miles." He may have exaggerated a bit, but the laws of physics do predict that a well-struck ball on the Moon could travel well over a mile (1,760 yards) in the rarified atmosphere and reduced gravity.

The golf balls naturally had the same mass on the Moon as they did on Earth and would have anywhere in the universe. Weight, on the other hand, being the force with which mass is pulled toward the center of the attracting body, varies from planet to planet. On the Moon, the pull of gravity and so the weight of the ball there is only about a sixth of what it is on Earth. Because Shepard's muscle strength presumably remained essentially what it was on Earth, any golf ball struck by him with the same force as on Earth would necessarily travel farther on the Moon.

Physicists tend to think in terms of mass, engineers in terms of weight. This is because traditionally engineers have focused almost exclusively on how objects respond to terrestrial forces, and as long as their structures and machines are built and operated near the surface of the Earth, weight is effectively as invariant as mass. In a skyscraper, the weight of a girder lifted off a truck at ground level is essentially the same as that same girder installed on an uppermost floor. However, when engineers are asked to design space capsules, Moon landers, and Mars rovers, they do have to take into account the different weights that the devices will have before launch and after touchdown—or work exclusively with mass in their calculations of strength and behavior. Either way, they have the advantage of designing devices that would not be strong enough to work on Earth but will function perfectly in a reduced gravitational field. This is why the Lunar Module that Neil Armstrong landed on the Moon looked so fragile and flimsy by Earth standards. Had engineers not been able to design such a light lander, the Apollo program might not have met President John F. Kennedy's goal of landing a person on the Moon within the decade of the 1960s.

In our everyday lives we have tended to ignore the distinction between weight and mass. In the United States, we order a turkey by specifying how much it should weigh in the traditional units of pounds and ounces. Just about anywhere else on the globe a shopper would specify a quantity of meat in kilograms. For milk, juice, and other liquid foods, we specify quantity by volume. We Americans became so comfortable using our familiar measures of things we eat and drink that we are likely to be thrown off guard if someone were to talk about, say, a pound of milk. This happened to me once, when my doctor was explaining what it means to be about eight pounds overweight. He likened it to carrying around a gallon jug of milk all day. It took me a moment to recalibrate how I think of a plastic container full of white liquid.

Each of the departments of a supermarket provides an excellent laboratory for getting a feel for and testing our sense of amounts, quantities, measures, and forces associated with foods packaged and unpackaged. In fact, going to a supermarket and making choices there involves not just judging mass and weight but using all five of our senses. As soon as we walk up to today's typical food mart, we hear the *sound* of automated doors opening for us, welcoming us inside, where we are immediately enticed by the *smell* of fresh fruits and vegetables and the *sight* of their colorful arrangements. The brazen among us may pluck a grape off a bunch and give it a *taste* test. And of course, we must *touch* everything we put into our shopping basket. But it is not strictly touch and take, the way touch and move applies in a chess match. We can pick up as many avocados or peaches as we like before selecting one to buy, and we can turn to the label on whatever can of soup or box of cereal we wish to read to compare in-gredients and nutritional value.

It was not always thus, for as recently as the middle of the twentieth century in the United States it was not uncommon for a shopper to visit physically distinct stores for vegetables, meat, fish, baked goods, and what need you. Shoppers walked along and crisscrossed city streets or negoti-ated a crowded farmers' market to visit each specialist merchant in turn, and in many cases they did not handle the goods; products were selected, wrapped, and handed to them by a clerk.

The concept of a self-service supermarket is credited to Clarence Saun-ders, who left school at age fourteen to work as a clerk in a general store. He eventually became a wholesale grocery salesman in Memphis, Tennes-see, and decried the lack of efficiency that existed in retail stores. Among the principal inefficiencies Saunders perceived was the use of clerks to re-trieve items, so he devised a system by which the customer would do that directly. Saunders's concept was to divide a store's floor plan into three distinct areas: (1) a front "lobby (forming an entrance and exit room),"
what today we might call a vestibule where shopping carts are picked up and returned and where checkout counters are located; (2) a "sales de-partment," which was entered through one gate or turnstile and exited through another and within which customers were constrained to move on a prescribed path and were thus "required to review the entire assort-ment of goods carried in stock," which were displayed on shelves and in appropriate cabinets, including a glass-doored refrigerator for perishables; and (3) a rear "storage or stock room." Above the shelving and cabinet units delineating the sales aisles were galleries, from which a "floor-walker" or other employee could direct and instruct customers in the new method of shopping, monitor the shelves and cabinets for need of replenishment, and generally supervise the activity below—all without getting in the way of the customers inspecting and selecting goods as they negotiate the maze. Saunders opened his first store to operate under the new principle in 1916 in Memphis. His patent application for a "self-serving store" reported that "the sales of a store thus equipped have exceeded, by three or four times, the amount of sales that it would be possible to handle in the same store waited upon by clerks in the usual way."

Although Saunders justified his invention largely by its economic advan-

A drawing from the 1917 patent for a "self-serving store" invented by Tennessee native Clarence Saunders, which put his concept in perspective. In the previous year Saunders had opened the first in his chain of Piggly Wiggly self-service grocery stores; the patent describes and documents the economic advantage of not needing clerks to retrieve individually each item on a shopper's grocery list. *From U.S. Patent No. 1,242,872.*

tage to the store owner, the self-service concept also offered advantages to the customer, who although constrained physically was not hampered sensually. By following a designated path through stocked shelves, the shopper came face to face with all the store's offerings, thereby being reminded of an item inadvertently left off the shopping list for the evening's dinner. Saunders expanded his concept into a chain of stores, each designated by the copyrighted and trademarked name Piggly Wiggly.

Today, the chain's website states that its inventor "wanted and found a name that would be talked about and remembered," though it admits that he was "curiously reluctant to explain" the origin of his choice. According to one story, Saunders "once saw from a train several piglets trying to get under a fence." Since a contemporary dictionary defined "piggy-

A circa 1918 photograph of the original Piggly Wiggly store, which opened in 1916 in Memphis, Tennessee, shows the popularity of the self-service concept that evolved into the modern supermarket. Note how the layout of the actual store differs from that in the patent drawing, which nevertheless did capture the essential features. *Library of Congress, Prints & Photographs Division, LC-USZ62-25665.*

wiggy" as a "child's riming extension of *piggy,* used playfully in speaking of or to a child," the constrained movement through the sales department of Saunders's store may have invoked from him the recollection, hence the name. Another explanation may or may not be Saunders's own reply to the question of why he named his chain what he had: "So people will ask that very question." Or perhaps he was simply more truthful than playful and named the stores after the way hungry shoppers had to wiggle their way through the crowded aisles and checkout lanes.

The now-standard layout of a supermarket in parallel aisles open at both ends allows shoppers to move freely from section to section of the store, following their noses and other senses, as it were. Displaying bread and milk in the farthest corner of the store from its entrance and produce section is not a mere accident, nor is placing the dairy case as far as possible from the bakery counter. By doing so, the store forces the shopper coming in only for a loaf of bread or a quart of milk to pass by many other potential purchases. Caveat emptor.

In a supermarket, all our senses should be on high alert. A Sunday magazine article on cooking cabbage in imaginative ways advised shoppers to "look for tightly packed heads; they should feel heavy for their size." We can compare heads of cabbage and lettuce, and pieces of produce generally, by holding like-looking ones in our hands and comparing them, using the inner sense of heft that we develop over years of shopping. Two equally sized melons that do not seem to weigh the same present a clue as to what might be inside. The lighter one will most likely have a greater hollow, something that might be preferable in a geode but not necessarily in a cantaloupe. Savvy shoppers will also knock their knuckles on a melon, sounding it for solidity. They choose peaches by squeezing them gently, to feel if the flesh is ready. Choosing among other fruits and vegetables, such as apples and bananas, may involve more sight and smell than sound, looking for the telltale blemish that suggests the fruit inside is overripe or even rotten. Too solid an avocado might not be ripe enough for dinner; too squishy a one might be past its prime. Shopping, like life, is a balancing act.

The frequent shopper develops a feel for the weight of an ordinary-size banana and thus can make a pretty good estimate of its weight and the weight of the bunch without using a scale. Nevertheless, it used to be the case in supermarkets that there were plenty hung throughout the produce department so that infrequent banana shoppers could readily check their feel for how much a bunch weighed, and therefore cost, before putting the bananas in the shopping cart. The old analog scales were rather simple devices and in their most basic and transparent form consisted of a pan hung from a spring hung from a hook anchored in the ceiling. Putting bananas in the pan stretched the spring, which caused an index or dial attached to it to indicate the weight on a linear or circular gauge. For fruit of about equal size, two bananas stretched the spring twice as much as one did. The mechanical principle that makes such a scale work is known as Hooke's law.

From 1662 to his death in 1703, Robert Hooke was curator of experiments at the newly established Royal Society in London. In his position he had access to a wide range of apparatus and was able to work in many areas of science and engineering, including but not limited to microscopy, materials science, strength of materials, and structural mechanics. Even before he joined the Royal Society, among Hooke's goals as an engineer was developing an accurate and reliable clock for ships at sea, for it was by comparing the local time to that in Greenwich, England, that a ship's longitude could be established. Naturally, one of the key components of any clock was its mainspring, and so Hooke experimented with springs. It was through this work that he came to discover the law of elasticity, which states that a spring's extension is proportional to the force with which it is stretched. He was reluctant to share his discovery, because knowing it would enable competing clockmakers to match the accuracy of his chronometer. At the same time, he wanted to establish that he was the discoverer of the law, and so he published it as an anagram, which he could later unscramble should he have to prove his claim. The anagram read, in the alphabetically ordered letters of the scrambled Latin words, *ceiiinossttuv*, which decoded became *ut tensio, sic vis*, and which translated states, "As the extension, so the force." In symbols, the law for a so-called linear spring is expressed as $F = -kx$, where the force F and extension x are related through the *spring constant k*. (It is called linear because the equation when plotted on *F-x* axes takes the form of a straight line; a plot of force versus extension data for a spring with constant k will follow a straight line with that slope.) The minus sign indicates that if the spring is stretched in one direction (say, in which x is positive), then the force resisting the pull will be negative, or in the opposite direction. This is consistent with our experience with common springs, which unless overstretched (and no longer elastic) will always seek to revert to their natural length.

It is easy to develop a feel for Hooke's law by experimenting with a rubber band looped around both thumbs. The more we move them apart, the more the rubber band is stretched, and the more it is stretched, the harder it becomes to stretch it farther, precisely because the force that it pulls back with increases with the extension. Naturally, if the rubber band is stretched too much, it will break. Then, the energy stored in the springy structure will be transformed into the sound energy announcing the break

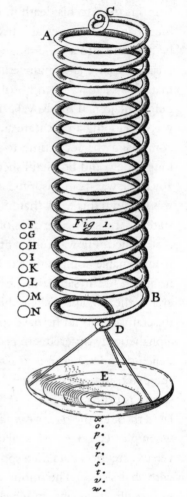

Robert Hooke, a contemporary of Isaac Newton, employed careful experimental observation and measurement to establish the existence of a linear relationship between the extension of a spring and the force needed to stretch it that amount. The relationship came to be known as Hooke's law. *From Hooke et al.,* Lectiones Cutlerianae, 1679.

and the kinetic energy associated with the strip of rubber flying away, possibly into the experimenter's eye. This is why protective eyewear should always be worn when performing experiments, no matter how simple they may seem.

Of course, springs are not only stretched, and the equation expressing Hooke's law works also when a spring is compressed. In this case, the extension is interpreted to be a negative number, which the equation's minus sign turns into a positive force, meaning one that pushes back against whatever is producing the compression. A typical retractable ballpoint pen can serve for an easy experiment. Pushing down on the button on the top of the pen compresses the spring inside. If we push the button slowly and attentively, we can feel the effect of the spring pushing back on our thumb, and with a noticeably increasing force as we continue to push down— until the ratchet or cam follower mechanism inside (different pen manufacturers use different means to the same end) clicks the cartridge into place, exposing the writing tip at the other end. When we are finished writing with the pen, we press down again on the button until the ratchet is released and the ballpoint cartridge pops back into the body of the pen. It is the energy stored in the spring when we deployed the cartridge that now propels it back as the freed spring returns to a less compressed length, ready for the cycle to be repeated. If in retracting the ballpoint we resist clicking the button quickly and hold our thumb on it to feel the force of the spring, we will sense it getting weaker as the spring gets longer. It typically does not reach its full uncompressed length because some residual compression is needed to provide the force that keeps the top button properly positioned and the ink cartridge from rattling. Regardless, what we feel as the ballpoint is extended and retracted is a physical manifestation of Hooke's law.

Not all springs look like a band of rubber or the helix of wire inside a retractable ballpoint pen. The top of a push-down-and-turn child-resistant plastic medicine container works because its top contains an insert that functions as a spring. In one version, when the cap is pushed down and turned clockwise to close it, the insert is compressed and held in that position by luglike projections engaging slots around the edge of the container proper. The compressed-insert spring exerts a force against the rim of the container and so keeps the cap pushed up and locked in place until someone pushes down on the top (thus compressing the spring even more) to separate the lugs and slots and, simultaneously, rotates it counterclockwise to disengage the lugs from the slots so that the top can be removed.

Large lessons lie in small things. No matter what shape a paper clip may take, it is essentially a spring. The so-called binder clip keeps together sheaves of paper that are too thick for a regular bent-wire clip; the clever device was invented by a Washington, D.C., teenager named Louis E. Baltzley, who came from a family of inventors. His grandfather was Elias

Howe, who is remembered for the sewing machine; his father and uncle also held patents. Young Louis, who would go on to invent small items including an easily picked up and stacked poker chip, a sifter top for powder containers, and a drinking glass holder for the side of a game table, wished to help his father, whose main occupation was writing, keep his manuscripts in order.

The binder gadget was patented in 1915 under the title "paper-binding clip," and throughout the subsequent century its appearance changed hardly at all from that shown in the drawings for Baltzley's patent, attesting to the difficulty of improving upon it. Its basic element is a strip of spring steel formed into a shape resembling a pup tent. Two steel-wire handles bent into a keyhole shape fold over the tent sides to serve as levers by which to spread the top of the tent open wide enough to receive a sheaf of papers. When the handles are released on the spring-steel tent, which wants to assume its naturally closed configuration, it clamps down on the papers so firmly that they do not easily slip out.

Squeezing the shapely lever-handles of a binder clip between the thumb and index finger provides a strong sensation of the springiness. It is easy to open up a little gap but increasingly difficult to open up a wider one, just as Hooke's law predicts. The clever design of the bent-wire handles enables the fingers to maintain their grip as they press with increasing force, which the user cannot help but feel. It is an ingenious device, and using it is an excellent way to feel the strong force of resistance that can come from even a small and compact device that looks nothing like a coil spring or the iconic Gem paper clip. It is also wonderfully adaptive. The handle loops allow the tightly clasped group of papers to be hung on a hook for ready access. When the handles are folded over onto the papers, they are out of the way for less visible storage. Alternately, the handles, which themselves are springs, can be manipulated to be removed entirely from the clip proper by squeezing their legs together sideways and angling them out of the cleverly formed cylindrical recesses into which the edges of the clip terminate. By thus removing the handles from the clip or clips holding a set of papers together, a virtually permanent binding in book form can be achieved, with the backs of the clips forming a spine of sorts. In fact, a label could be affixed to the nearly flat back of a clip to identify what it holds.

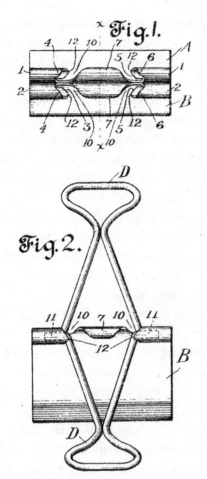

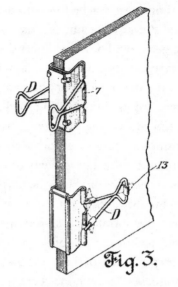

The binder clip, which can hold together a much thicker stack of papers than a conventional bent-wire paper clip, was patented in 1915 by the teenage inventor Louis Baltzley. It consists of two spring elements: the clip proper (*B in "Fig. 2"*) and the removable handles (*D in "Fig. 2"*) that serve as levers to open the clip wide enough to bind a thick sheaf of papers (*"Fig. 3"*). *From U.S. Patent No. 1,139,627.*

It was the absence of a spring function that hampered the development of eyeglasses as we know them. Spectacles are believed to date from the late thirteenth century, when lenses were mounted in frames made of anything from bone to metal to leather. A mounted lens generally took the form of a small magnifying glass. When two such handheld glasses were connected by means of a rivet through the base of their short handles, the resulting pair could be balanced on the bridge of the nose for reading and other close work. Unfortunately, users had to hold their head somewhat back and rather still lest the spectacles fall off. This led to the introduction of a variety of appurtenances for keeping the set of lenses in place, including extensions that reached up and across the forehead and were anchored behind the head or under a hat or cap. Strings attached to the sides of the glasses could be looped over the ears; the Chinese employed weighted strings that draped behind the ears. Early temple pieces ended at the cheekbones and depended upon spring action and friction to hold them in place.

The introduction of spring steel for a bridge piece between two framed lenses produced the pince-nez, which literally pinched the nose to stay in position. These spectacles were not worn high on the nose, where it is hard and bony, but low where the fleshy part can be compressed into the nostrils. This did keep the glasses in place, but at the expense of easy breathing. Modern eyeglasses typically have more rigid bridges and often incorporate unobtrusive springs into the temple pieces, which keeps them pressed against the side of the head for a snug fit. Some temple pieces rely on the springiness of the frame material itself to provide a desirable degree of snugness. Plastic frames became fashionable in the 1950s and have remained popular. When new, the springiness of the plastic allows the glasses to fit snugly, but in time the plastic will yield to the constant force of the relatively uncompressible human temples relentlessly pushing the temple pieces apart. This causes the bridge to be become permanently bent outward, which widens the space between the temple pieces and allows the glasses to slip easily down the nose. This phenomenon of a constant force causing a structure to be reshaped is known to engineers as creep. Making the plastic thicker and heavier only postpones the inevitable. People who have worn a pair of plastic-frame glasses for some time have constantly to push them up along the ridge of the nose with a motion that often becomes a tic.

For me, lightweight metal frames with spring-loaded temple pieces were a godsend. When properly fitted, the frames lightly caress the side of my head with a gentle force that I can feel only enough to reassure me that the glasses are not going to slip off when I bend down to tie my shoe or swing my head quickly to see what startled me. But wearing these eyeglasses day after day meant that the temple pieces were constantly pressing inward on the side of my head. Since metal is generally less subject to creep than is plastic, the push is unrelenting and leaves a distinct depression in the hair above my sideburns. Even small forces applied over a long time can leave their mark as surely as a river running for eons can carve out a grand canyon.

Because they so frequently move out of position, spectacle frames have been described by ophthalmologist Melvin Rubin as "one of technology's best examples of poor engineering design." Still, wearing a pair of glasses provides a veritable laboratory for feeling forces even if experiencing the same ones every day becomes so familiar a sensation that we notice them only in their absence. This can be the case with lifelong eyeglasses wearers who have cataract surgery to replace their natural lenses with artificial ones, thereby obviating the need to wear glasses. With the spring forces exerted by bridge pads and side pieces no longer pressing on the nose and temples, the absence of the glasses and their attendant forces becomes apparent. Longtime glasses wearers may find themselves instinctively reaching to push their absent spectacles up the bridge of their nose. Just as amputees are said to continue to feel the sensation of a missing limb, so eyeglasses wearers separated from their glasses frames can imagine that they still feel them in place.

A trampoline is a spring, and so is a tennis racket. A tennis ball and anything that bounces can also be thought of as a spring. To someone watching a tennis match, it may not look as if the ball is compressed when hit by the racket, but high-speed photography captures the fact. During the impact, the strings of the racket get stretched and the ball gets flattened where the contact occurs. The racket strings, which are springs in their own right, want to return to their normally taut position, and in so doing they act like a phalanx of slingshots giving pushback to the ball; the ball, wanting to return to its round position, exerts an opposite force on the racquet. These forces of restoration effectively add to the force the swinging racket imparts to the ball. In combination, all these forces drive the ball toward and, with any luck, over the net while it stays in play.

Conventional helical springs are generally intended to produce an axial force, such as the one that propels a ballpoint back inside the pen barrel, but the actual value of the force needed to make the pen work is generally of little interest to anyone but the engineer designing the writing implement. Similarly with a tennis ball: it is not so much what amount of force is involved as how effectively the ball bounces off the racket and the court surface. Engineers measure this not by a spring constant but by a coefficient of restitution, which is the ratio of the speed of the ball leaving a collision with a relatively immovable object such as a tennis court to the speed of the ball approaching it. If a ball dropped from a certain height sticks to the ground like a blob of putty, the coefficient of restitution is 0. If the ball bounces back up to the height from which it was dropped, the coefficient of restitution of the ball-ground pair is 1. Neither extreme is achievable with a real ball and a real ground surface, for virtually all materials have some springiness to them and some energy is always lost in the form of sound, if nothing else. If the coefficient of restitution is between 0 and 1, with each subsequent bounce the ball will rebound to a lesser height, and eventually stop bouncing altogether. Generally speaking, the goal of both sports equipment manufacturers and their customers is to make balls and the implements that strike them have as high as possible a coefficient of restitution.

Because everything has a certain amount of springiness to it, just about anything can be considered a spring in its own right. A diving board, whose cantilevered end will deflect in proportion to the weight standing on it, is a spring. A diver, by timing preparatory bounces with the natural frequency of the board, can get an assist from it as it pushes back up after being pushed down. More complex arrangements of beams and girders, such as in tall buildings, also possess a springiness. A gust of wind pushes a building sideways, but it springs back, oscillates, and attenuates as the gust passes.

The legendary dining hall food fight depends upon springlike action. A glob of mashed potatoes in the bowl of a spoon can be flipped across the cafeteria because the system consisting of the eating utensil and the gripping hand of a mischief maker taken together possess a kind of elasticity. Unless very cheap silverware or an illusionist such as Uri Geller is involved, the utensil cannot be bent very easily, but in combination with the way it is held the system becomes a spring. This can be demonstrated by holding a spoon upright in one hand and pulling back on its bowl with a finger of the other hand. When properly held and released, the spoon will spring forward and stop suddenly when it reaches a near upright position. The mass of potatoes, only loosely stuck to the end of the spoon, will separate and continue on its way by means of its inertia. The outside force that will stop the projectile is provided by the target's forehead. That force, albeit reacting to a small amount of soft and fluffy stuff, will be magnified by feelings of embarrassment and anger and will likely be retaliated against with appropriate counterforce. For every action there is an equally apposite reaction.

Let's return to the supermarket and imagine ourselves moving up and down its aisles and taking cans, boxes, and bags of food from the shelves and placing them in a shopping basket. If it is a handheld basket, the weight and possible imbalance of just a few heavy items can make us sorry we did not take a wheeled cart. But as a cart accumulates groceries, we would feel it getting harder to push and steer. The accumulation of weight naturally bears down harder on the wheels and they on the floor. Without the friction that develops when we push the cart, its wheels will not rotate, for there would be no tangential force to produce rotation. Just as we cannot walk on a frictionless surface, so a wheel cannot roll on one. Tough industrial carpeting can provide sufficient friction to turn the wheels of the most heavily loaded shopping cart, but it can also produce resistance to wheels rolling over it. This kind of resistance is often referred to as rolling friction, but in fact it is due to a lot more than friction. Much of what resists the cart's progress on a carpet is actually the little hill that develops in front of a wheel when it depresses the surface, and the softer the surface, the higher the hill and hence the greater the resistance to the wheel rolling over it. Concrete, terrazzo, and other very hard surfaces naturally minimize rolling friction. Supermarket designers have to choose between a hard floor, onto which a dropped jar of pickles is sure to break and leave a mess, and a more forgiving floor that can absorb the impact but across which it will be a bit more difficult for a shopper to push a cart.

Wheeled luggage can be pushed or pulled. When air travelers lead a two-wheeled rollaboard, their hand pulls up and forward on the handle, and any small bump, lip, or other irregularity along the concourse is simply rolled over. However, when the luggage leads, their hand imparts to it a forward and downward force. This latter component effectively adds to the weight of the bag, which increases the rolling friction and pushes the wheels into a bump or void rather than pulls them over it. Anyone who has tried to push a roll-on onto an elevator, thinking that it will be positioned for a straightforward exit, has likely experienced the wheels getting stuck in the gap between floor and car. When the bag is pulled onto the elevator, however, that gap is easily negotiated. The traveler feels the bump but can pull the bag over it without missing a step. How a force is applied to an object generally matters much more than what the object is.

Eggs are notorious for being broken in transit from supermarket to home. Even when the shopper has opened a carton at the store and checked that all the eggs are intact before adding the lot to the shopping cart, cracked ones can be discovered when they are being put into the refrigerator's egg caddy. Clever packaging is obviously available to protect the fragile cargo from destructive forces, but no protection can be guaranteed to be totally effective. Fragile potato chips are packaged loose in air-tight bags that feel like inflated pillows, and they do a pretty good job of keeping the chips intact. We know from experience how easy it is to snap a potato chip or crack an egg, and so we may marvel that so many do survive unbroken after the multiple times a bag or carton is handled from when first packed to when unpacked at home. Effective packaging is definitely something to be admired. But what works for one item may not for another: imagine loose eggs in a potato chip bag.

Many engineering students are introduced to the realities of packaging through the good-natured tradition of the egg drop contest. The idea is simple: design and build a device to transport a raw egg from the roof of the engineering building to the concrete walkway leading up to it. The winner will be the lightest device in which the egg survives unbroken. In other words, the egg must descend slowly enough so that the force of impact does not crack the shell. Entrants must conform to a few simple rules, such as a maximum size or weight of the device, that do not inhibit creativity. Nevertheless, many entries tend to be but variations on the parachute and shock absorber.

One year at Duke, an environmental engineering student looked to nature for inspiration and found it in the samaras that appear on maple trees each autumn. As these winged seedpods fall from the tree to the ground, they twirl gracefully and land their cargo softly. The clever student used ordinary cardboard to construct a scaled-up version of a single-winged samara. In place of the seedpod he placed an exposed egg, held firmly in place by the friction forces between the egg and the edge of the ovaline void he had created in the cardboard. Because the faux samara was traveling so slowly when it hit the ground, the impact force was indeed small. The egg survived whole and remained in place. The crowd of onlookers spontaneously erupted in applause, and the student's winning entry was put on display in his department's trophy case. An egg drop competition has become a common challenge for grade and high school students enrolled in science, technology, engineering, and mathematics (STEM) programs and curricula that promote an integrated approach to the interrelated disciplines. Students are never too young to appreciate how understanding a concept like force transcends a single category of learning.

13

A Round Cake in a Square Box

And a Sagging Triangle of Pie

In the BBC television series *The Crown*, the free-spirited Princess Diana is depicted as being noncompliant with the staid restrictions placed on members of the royal family. In one episode, Queen Elizabeth II and other members of the Firm are discussing whether Lady Di will ever behave like one of them. "In time, she will give up her fight and bend, as they all do," the Queen Mother proclaims. "And if she doesn't bend, what then?" asks Queen Elizabeth, to which Princess Margaret responds, "She will break."

All things bend under pressure or stress (either of which in a mechanical context is nothing but another name for intensity of force). Someone or something that bends too easily ceases to be wholly itself; it becomes a floppy version of what it was meant to be. A person with too stiff a backbone that does not allow for adjustment to circumstances cannot easily function in society.

Human patience can be tested waiting for something to happen. One commonplace metaphor for being under pressure, especially in an office environment, is breaking a pencil in two. Theoretically this can be accomplished in a variety of ways, but two stand out. The one-hand method is to grasp the pencil in a tight fist and push against it sideways with the thumb. It takes an uncommonly strong thumb to break the pencil because the distance between the pushing and resisting forces is too small to give much of a mechanical advantage. The second method is to hold the pencil with both hands, grasping it as if it were the handlebars of a trail

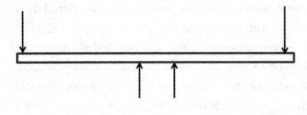

A pencil held in two hands can be broken by using the thumbs to push against its middle at the same time that the fingers pull its ends in the opposite direction. *PowerPoint drawing.*

bike. With the pinky of one hand near the point end of the pencil and the other near the eraser, the thumbs push out on the middle section of the pencil while the fingers pull in on the ends. Because there is a good distance between thumb and pinky forces, breaking will occur relatively easily, and the seemingly universal expression of anger, frustration, and rage will have been achieved with more flair than effort.

The actions of bending and breaking correspond to the main criteria engineers take into account when designing anything from a small submarine for plumbing the deep ocean to a tall rocket for launching missions into outer space. Each must be sufficiently strong to resist breaking under the forces to which it will be subjected, and each must be sufficiently stiff to maintain its shape under those same forces. Some things, such as a fishing rod capable of landing a marlin, are strong without being very stiff. Others can be stiff without being overly strong. A piece of crystal stemware can be shattered to pieces by the invisible pressure wave of a single high vocal note, something that may be done deliberately to demonstrate what can happen when the natural frequency of a vulnerable object is matched by that of a pressure wave emanating from a high-pitched vibrato. The resonance amplifies the effect of the force impinging on the bowl of the glass, and disaster follows. Other forces can break a glass, of course, and these obviously include accidentally dropping it on a hard surface or letting it bang too forcefully against abutting glasses in the dishwasher. That is why good crystal must be hand washed, which is the way Catherine takes care of the delicate stemware one of her aunts had shipped to her from Sweden as a wedding present.

Catherine had set out this crystal for a holiday dinner at which we were entertaining a colleague of mine and his family. We adults were enjoying after-dinner drinks and conversation when my colleague's wife offered their curious five-year-old a sip of water. She held the glass by its stem and put the bowl softly to his lips. He promptly bit into the rim and shattered the bowl as if it were the shell of an egg. The act of destruction was excused, but Catherine no longer was able to set a formal table for twelve. She did order a replacement wine glass from Sweden, but it was just not as thin and delicate as the one that had been broken. Evidently, in the intervening years the pattern had been made more robust, perhaps in response to its having been too easily broken. Yet not a single piece of Catherine's crystalware had been damaged by an adult putting it to his lips, drinking from it, and returning it to the table. Should an elegant design have to be redesigned in anticipation of its being used inappropriately?

Engineers design a critical piece of infrastructure to endure forces expected under normal circumstances as well as those reasonable ones expected to be encountered only rarely. Thus, a bridge will be designed to withstand day-to-day traffic and normal winds as well as the occasional overloading due to a traffic jam or winter storm. The day-to-day traffic load, plus a safety factor, will be taken as the baseline for sizing the components of the structure, which will determine its strength and stiffness. The forces associated with a low-probability event will also be considered, but in the absence of government requirements it will be a judgment call about whether the extreme forces considered will be those of a storm that is expected to occur only once every century or two.

Engineers understand the many manifestations of force and use this knowledge to design all kinds of structures, machines, and systems that give us shelter, convenience, pleasure, and joy. It is by appreciating beforehand what forces will act on a bridge or a skyscraper or a roller coaster—and how they will be resisted—that engineers establish its safety while it is still a concept on paper or computer screen. And even though a structure may be deemed strong enough to stand up to the force of the wind, if it flaps, bends, twists, or sways too much in a storm, the people using or occupying it may become uncomfortable, to say the least. They may actually fear for their lives on a wobbly bridge or in a swaying building. A well-designed structure must not only be strong enough to resist any forces threatening to crush, pull, or tear it apart but also be stiff enough to maintain a semblance of its shape.

Nature appears to follow a similar design philosophy. The shell of an egg must be strong enough to hold up against the outside pressure of the tube through which it travels on its way to being laid, as well as against the weight of the hen that sits on it to hatch it. But if the shell were too strong and stiff, the chick developing inside might not be able to exert sufficient force on the shell to bend and crack and ultimately break it. Strength and stiffness in nature can also be negative qualities. A kidney stone may survive intact its passage through a patient's ureter and urethra, but the effect can be excruciating. At the same time, the stone can be shattered into small and more easily passed pieces by the shock waves employed in the noninvasive medical procedure known as lithotripsy. Questions of how strong is strong enough and how stiff is stiff enough are ubiquitous in nature and in manufacture. Strength and stiffness must be combined in a balanced way to produce an efficient structure.

The steel shipping container has changed the nature of international trade and commerce, but the packaging and lading of ship cargo has always involved compromise. Crates, barrels, and sacks had to be strong enough to be manhandled by stevedores but not so strong as to be too heavy to be moved safely and economically. They had to be able to take the abuse of being loaded and unloaded at docks around the world, not to mention be able to survive the forces imposed on them in a rough sea. When in position upon a ship, they had to bear up under the burden of like things stacked upon them. Similarly, the twenty-foot-long steel-box container must be capable of supporting many like ones piled like building blocks upon it. No matter how bumped or dropped, each box must also be able to retain its shape so that it can be properly engaged by cranes and remain properly interlocked with its neighbors during transit and storage. These objectives are now achieved by using a sufficient amount of steel in forms that resist breaking, bending, and buckling.

The internationally uniform shipping container is an enormously successful design, but so are more modest-sized boxes made of much less strong and stiff materials. I know this from personal, if vicarious, experience. On Saturdays I often accompanied my father to the bakery shop, where my mind found the actions of the clerk behind the counter as interesting as the confections in the display case were tempting. My appetite for understanding the objects, devices, and gadgets being employed in the aromatic store was insatiable. I was fascinated by the machine that sliced a full loaf of bread in one pass. After the clerk lifted the segmented loaf from the machine with two hands, holding it as if it were a concertina, she upended it and balanced it in the palm of one hand, freeing the other to fetch from under the counter a flat white paper bag that she opened midair with a single deliberate motion accompanied by the sound of a whip snapping. What she had done was snare the air, which in turn pushed against the bag's inward folds, which billowed out like ship's sails. With the wide open bag looking like a chef's hat atop the head of bread, she simultaneously pushed the loaf up into the bag and pulled the bag down onto it. The clearance between contents and container appeared to be as tight as a piston in a cylinder. She left the bag open, lest the fresh bread be crushed in the course of closing it. She did not have to be so careful with a loaf of rye bread, because its crust gave it both stiffness and strength to resist. Watching the whole seamless process and imagining the forces involved was to me the greatest thing since sliced bread itself.

Another operation that captured my attention was the act of placing a pie or cake in a box, which itself had first to be formed from one of the large, flat, irregularly shaped sheets of paperboard stacked beneath the counter. The sheet had been deliberately die-cut to have an outline that suggested it was a piece of a giant jigsaw puzzle. Seasoned shoppers looked for other sweets to buy, but I was an audience of one rapt by the trick the magician was performing before my eyes: The bakery clerk folded three sides of the latent box up and joined them together by a tab-in-slot sleight of hand. Into this partially formed cul-de-sac of a box she partly lowered from the top and partly slid in from the front the freshly baked pie or cake my father had selected, displacing neither a crumb of crust nor a swirl of icing in the process. With the dessert in place, the last of the box's sides was raised as if it were the tailgate of a delivery truck and made secure by another tab-in-slot maneuver. As the shop woman swung the cardboard top down to close the box, her deft hands kept the flaps from invading the space between the inside walls and the contents by using her thumbs to guide the flaps outside, where they would be free to flutter like the wings of a dove had she not secured the box with cotton baker's twine, sometimes plain white but often candy-striped red and white. The medley of forces that conjured a treat box out of a flat sheet constituted an act of true legerdemain.

The baker's twine was kept on large spools in the shape of truncated cones. These were free to turn on strategically located spindles. I know this because they were in plain view, either sitting directly on the work counter or suspended right above it. But no matter from where it was deployed, the twine played out easily, if a little noisily, as it was wrapped around the box seemingly with the speed that Superman girdles the Earth. First three or four turns secured the top and side flaps, and then three or four crosswise did the same for the front flap. In preparation for tying the finishing knots, the clerk cut the twine with the flick of the finger on which she wore a pragmatic stainless-steel ring set not with a stone but with a small knife blade shaped like a miniature sickle. When two or more boxes were to be carried home, they were stacked in a tower formation and tied together with more string.

I gladly accepted the challenge of carrying the saccharine burden home. As tightly as the twine had been wound around it, there was always a little slack in the assembly resulting from the elasticity of the twine itself and the flexibility of the thin paperboard boxes. This not only allowed me to get my fingers between the twine and the top of the upper box but also allowed a degree of springiness so the up-and-down movement of the stack resonated with the bounce in my step. Because the boxes had to be held up and away from my body, lest they drag on the concrete or bump against my leg, I necessarily had to hold my arm up and angled out somewhat, and this made for a tiring situation. Still, I felt a sense of responsibility, accomplishment, and pride in taming wild gravity with a whip of twine.

One thing did bother me, and that was that the bakery goods were usually round and the boxes square. But the world of things is full of awkward pairings: mismatched pegs and holes marked by misfit, imperfection, and compromise. Yet even in these shortcomings there are lessons to be learned about why and how they could not be avoided, about the forces behind the forms. And their stories reveal the secrets of how so many things do fit together, even if imperfectly.

The ways of the old-time bakeshop are absent in the baked-goods department of today's supermarket, where so many items are prepackaged in transparent plastic containers formed to conform to the shape of their contents. Round pies tempt us from beneath transparent round domes. Tops snap onto bottoms and lock in place without the use of twine or tape. The packages are neither biodegradable nor easily carried with one hand, but they do fit, if not geometrically congruently—along with the cylindrical cans and bottles and globular and tubular fruits and vegetables—into the ample trapezoidal hold of a shopping cart containing also prismatic boxes of cereal, pasta, and cake mix. The incongruity of things made and things found is a fact of life, and like most facts of life this is accepted as the way things are. Maybe it would be a more efficient use of natural resources and space if all things, including pegs and holes, packaged goods and shopping carts, were square and cubical, but who wants to slice a square pie?

Not only do we generally accept the shapes of things as we find them, but also we adapt to their idiosyncrasies. One example is the round pizza carried out or delivered in a squat, square corrugated cardboard box. Early pizza boxes were made of the same thin and unreinforced paperboard as bakery and cereal boxes, and they were hardly suitable for bearing the weight of a large pie distributed over a sixteen- or eighteen-inch-diameter circular area. Paperboard worked for bakery pies and cakes because they came in pie tins or on cake pallets that provided the stiffness that the box bottom did not have. Pizza pie boxes, on the other hand, must rely on their geometry alone to maintain the shape of even the notoriously thin, soft, and floppy hot contents of the New York style pie. If there is anything between the pizza and the box, it is a flimsy sheet of waxed paper, whose purpose is to keep the olive oil from soaking into the cardboard and jeopardizing the structural integrity of the box.

Still, into the middle of the twentieth century fresh-from-the-oven pizza pies were packaged in boxes made of material that worked fine for baked goods sold at room temperature but were far from ideal packaging for something just removed from a very hot oven. The high heat and moisture content of a fresh pizza made the large but shallow boxes holding them even flimsier, floppier, and saggier. Making a pizza box out of stronger and stiffer corrugated cardboard was a way to ameliorate this problem, and in the mid-1960s Domino's Pizza founder Tom Monaghan commissioned one to be designed and manufactured.

The corrugated cardboard pizza box gains rigidity by being closed. We know that a sealed box of cereal holds its shape and protects its contents through the successive processes of being filled at the factory, shipped to the supermarket, stocked on the shelf, placed in the cart, thrown into the grocery bag, and jostled in the station wagon before being placed intact on the breakfast table. As soon as the box is opened, however, we feel its rigidity drop because it was the sealed top that kept the sides in place.

The importance of geometry for a container can be demonstrated easily by reverse engineering a typical cereal box. The process begins with opening the top, emptying out all contents, and unsealing the bottom. This leaves an open-ended structure, a paperboard tunnel readily felt to be much less rigid than its closed counterpart. The doubly opened box can be sighted through, as if it were a tube. However, unlike a circular tube, whose shape cannot be appreciably changed without some crushing, the rectangular tube of the cereal box can be transformed into a parallelogram without altering the shape of any of its sides. It is this property of a rectangle that requires one or more diagonals to be present in a series of structural rectangles in the frame of a building or bridge. A diagonal subdivides a quadrilateral into triangles, which hold their shape, which is why triangles tend to dominate the structural pattern of so many bridges. One so characterized is known as a *truss*, a word that derives from the Middle English *trusse*, among whose meanings was "a bundle of sticks." Diagonals are also visible in the framing of steel buildings under construction but become concealed behind the architectural facade of a completed structure. Without diagonals or the bracing provided by structural details—akin to the top and bottom of a closed cereal box—a rectangle is more mechanism than structure. Finding the seam that typically runs down one corner of a cereal box and separating its glued surfaces allows the box to be deconstructed fully into a flat sheet of pasteboard not unlike one that the bakery-store clerk pulls from beneath the counter.

Sturdier boxes, such as those used in shipping heavier items such as laptops, printers, and other electronic equipment, are made of corrugated cardboard, which is stiffer and stronger than paperboard. Cardboard can even be employed to make designer tables, chairs, and other pieces of furniture, as the architect Frank Gehry began to do in the early 1970s. In 2021, dormitory rooms in the Tokyo Olympic village were outfitted with recyclable cardboard beds capable of supporting athletes weighing as much as 440 pounds.

The corrugated cardboard box is remarkably well suited for transporting a hot pizza, and this is no doubt why for decades it has remained generally unchanged in design and use. Starting out flat as it does, the box takes up relatively little storage space, which is often at a premium in mom-and-pop pizza shops. The corrugation process forms air channels in the sandwich structure, thus giving cardboard the insulating properties that help keep a boxed pizza warm. But nothing is perfect in design or use. The common pizza box certainly has its limitations, shortcomings, and downright failures, which tend to be excused and adjusted to by proprietors and customers alike. However, few flaws escape the eyes of inventors, designers, and engineers, who are always looking to improve on things. Yet even among these groups, the fact that a square box is used for a round pie is hardly worth further comment, other than to mention that there have been evolutionary changes that have addressed the problem—if it can even be called that—of the geometric mismatch. Domino's Pizza developed its polygonal box that better approximates the circular geometry of its contents—and incidentally distinguishes its boxes from those of its competitors. There are round pizza containers made of Styrofoam and other moldable materials, but such extravagant solutions seem out of proportion to the problem.

There are other than geometrical problems with the pie-in-a-box concept that do lend themselves to simple and inexpensive solutions. But these kinds of problems tend to be invisible to or ignored by all but the most persnickety of consumers and inventors. One such problem manifests itself when the top of a box containing a large hot pizza sags excessively because of the softening effect of the steamy enclosed environment. The problem can also arise when a boxed pizza is transported in a car that bounces over railroad tracks or potholes. After such a ride, more cheese may be found stuck to the underside of the box's top than remains on the pizza itself. Tolerant customers might scrape the cheese off and redistribute it on the pie, but inventors tend to be intolerant of everything but their own inventions.

Although water emptying from a bathtub tends to swirl clockwise down the drain in the northern hemisphere and counterclockwise in the southern, pizzas in a bouncing car are propelled straight upward everywhere on Earth. The Argentinian inventor Claudio Troglia figured out a way to keep pie and box apart and in 1974 received a patent for a "pizza separator," which took the form of a miniature stool with three legs that rested on the center of the pie. There is no mention of Troglia's device in a U.S. patent issued a decade later to Carmela Vitale of Dix Hills, New York, although her plastic tripod solved essentially the same problem, but from an upside down point of view. Instead of emphasizing the role of the pizza under the box top, her title "package saver" emphasized the box over the pie. Regardless, she was evidently bothered by the universal annoyance of the cardboard eating too much cheese.

As Vitale explained in her 1985 patent, cheaply made and disposable containers, "particularly those used to deliver pizza pies or large cakes or pies, comprise boxes with relatively large covers formed of inexpensive board material" that have a tendency "to sag or to be easily depressed at their center portions so that they may damage or mark the pies or cakes during storage or delivery." Her solution was "to provide a lightweight and inexpensive device" molded "from one of the plastics which is heat resistant such as the thermo set plastics and which will resist temperatures of as high as about 500°F." The device that she preferred and illustrated in the patent was a utilitarian tiny tripod that resembles a minimalist three-legged stool. In her description of the so-called preferred embodiment, she explained that the legs should have "a minimal cross section to minimize any marking of the protected article" and also "to minimize the volume of plastic required" and thereby keep the cost low.

The device has been described variously as a "lid support" and a "pizza stacker." One version of it is sold under the brand name Hercules, although Atlas might also be apt. Vitale's term "package saver" was belied by her claim that the true benefit of the tripod was in "preventing damage to the packaged food article by the cover." Thus it is not so much the disposable package that is being saved but its contents. The box-lid support device has since been appropriately renamed "pizza saver," though few people who admire the thing for so effectively saving their pizza from being scalped of its cheese know it by that name. In fact, it is most popu-

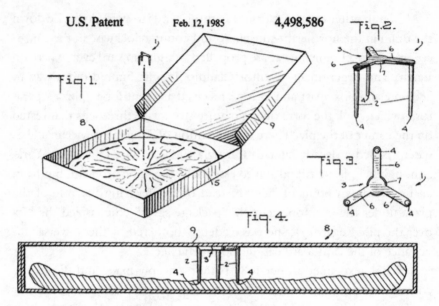

U.S. Patent   Feb. 12, 1985   4,498,586

Fig.1.   Fig.2.   Fig.3.   Fig.4.

Patent drawings illustrate how a small plastic tripod can be inserted between a pizza pie and the top of the box in which it is being transported, thereby keeping the two from coming into contact. *From U.S. Patent No. 4,498,586.*

larly referred to as an all-purpose "little thingy," which in context everyone understands and conjures up an image of the miniature tripod. People also appreciate that with its thin, spindly legs it has a very small footprint on the pizza and so consumes very little of the pie itself.

Although possessing no great grace or elegance of form, Vitale's answer to the problem of the cheese-consuming box top is considered an elegant solution technically. It has no unnecessary embellishments. Clearly, the small and inexpensive device was meant to be as disposable as the pizza box, but not everyone who first sees a pizza saver can throw it away. Persons engaged in arts and crafts activities especially tended to wash it off and put it aside for some future unspecified use that they were sure they would someday find for it. For some, the "pizza table" came in handy when a dollhouse room needed a piece of scaled-down furniture. One woman, whose hobby was decorating eggs, found that turning a pizza saver upside down made it into an ideal easel for holding an egg. She did have one complaint, however, and that was that the little thing was rather expensive, a fault that she attributed to having to pay for its elaborate packaging—namely, a large pizza in a larger box, which collectively cost her in excess of ten dollars.

As elegant a solution as the pizza saver might be, not everyone saw it to be without aesthetic, economic, and structural flaws of its own. One inventor identified a critical shortcoming—especially for savers with a solid round top: How was it to be packaged when purchased not one at a time but in bulk? The mom-and-pop pizzeria wanting a supply on hand might even buy them in quantities of a thousand, at a cost of a penny or so apiece. The way the savers were typically packed was to be thrown loose into a large cardboard box, whose contents settled during shipping the way cornflakes do. The half-empty box took up an inordinate amount of space in a cramped pizza shop. The inventor's improvement was to incorporate into the top of the saver holes into which the legs of a second one could be inserted. Savers nested in such a way would naturally take up less space than a random jumble of them. What the inventor seemed to ignore, however, was the labor that would be involved in nesting the tripods, thereby adding to the cost of something whose price was supposed to be kept to a minimum. Inventors of things large and small often lose sight of the bigger picture.

But you don't have to be a formal inventor with patents to your name to come up with a clever solution to an annoying problem. Once Catherine and I had a pizza delivered to our hotel room, and we admired the creativity of the pizzeria in finding a substitute for the tripod version of a pizza-saver that must certainly have saved storage space just by using something already at hand but less specialized. Instead of being set with a little table, the center of the pizza was occupied by an upside-down plastic cup of the kind used in fish fry and other informal restaurants to contain small servings of ketchup, tartar sauce, and other condiments. The elegant and nestable solution was clever indeed.

Some inventors, like still-life painters who crowd as much as possible onto a canvas, try to solve multiple problems with multipurpose devices replete with features. Sometimes their inventions succeed; sometimes they do not. In 2000, the Iowa inventor Mark Voves received a patent for the design of a "pizza cutting and eating tool," that is, a slicing wheel and fork in one, which would enable the fastidious pizza eater to cut and spear pieces of pizza without having to use a separate knife and fork. The one-handed method would also free up the diner's other hand to hold a smart phone and take a selfie. A team of inventors received a patent for what they described as a "lid support and serving aid," which was embodied in a plastic tripod with one of its legs inclined, elongated, and serrated so that it could be used to cut through fused cheese and crust to separate individual slices. The top platform of the tripod contained a hole large enough for an index finger so that a good grip could be had for executing the cutting process. The platform was shaped so that individual tripods could be nested like function-room chairs to save shipping and stocking space.

Not all inventions are patented. Those that are not may not even be thought of as inventions by their creators, who hardly think of themselves as inventors. Yet their ideas can become implemented so widely as to be termed innovations. Among these are small changes in how we carry out a common activity of daily life. These ideas certainly come from individuals, but they tend to diffuse so thoroughly through a community or region that who exactly was the inventor is virtually impossible to say. An example of this can be found in the practice of eating a sector of a circular pizza pie—a "slice."

Holding the slice by its crust makes it into a cantilever beam. This works well with a flatbread pizza, say, because its carbohydrate foundation tends to be dry and will hold its shape. A New York style pizza, on the other hand, tends to be soft and soggy, and a slice of it has a propensity to sag, drip olive oil, and shed a layer of cheese and tomato sauce. The slice can be served on a plate and eaten with knife and fork, of course, but when in 2014, Bill de Blasio, New York City's then-new mayor, did this during a political visit to a Staten Island pizzeria, he shocked local onlookers. His behavior was mocked, and according to one commentator, violated "the longstanding city protocol of devouring pizza, no matter how greasy, with the hands, and the hands only."

In fact, only one hand is used in the traditional New York method, which is to fold the slice along a radial line. This effectively creates a trough that, when the creased slice is held with the crust slightly lower than the tip, contains the toppings. The method works because folding any kind of structure increases its stiffness. The principle is at work in the dimpled paper plate, the pie tins and cake platters used in bakeries, and the cigarette paper used for roll-your-owns. And many food products gain stiffness by being formed with folds or bends, including Ruffles potato chips and Fritos Scoops! corn chips. The folded taco shell and fluted tortilla bowl exploit the same principle.

The structural concept can be easily and cleanly demonstrated with a single sheet of paper. Flat and unfolded, it sags readily when lifted by one edge; folded one or more times—as is done in the process of making a paper airplane or pleated fan or corrugated cardboard—the sheet gains considerable depth, which equates to a much greater stiffness and ability to hold its shape. Achieving stiffness by folding and otherwise out-of-plane shaping is a ubiquitous technique in structural engineering, being exhibited in everything from grooved, ridged, and fluted tin roofs to sculpted concrete shells to molded automobile panels. It is why the extendable steel tape measure is dished, why steel shipping containers have corrugated sides, and why fifty-five-gallon steel drums have a pair of raised rings. In the case of the drum, the rings not only serve to stiffen the sides of the drum against buckling and denting but also provide a projecting ledge that prevents a drum from slipping from the grip of a device used to lift and move it. In addition, the rings serve to maintain a gap between drums in compact storage so that the claws of any gripping or lifting device will have room to reach in between the juxtaposed drums. Whether intended or not, mechanical features seldom have only one or even just two purposes.

The everyday actions of carrying a stack of boxes or folding a slice of pizza are seldom recognized as the introductions to the world of forces and flexibility that they are. Yet the same forces that we feel acting on paper and crust are also ones that act on houses and cars, planets and stars. By feeling the forces that act so directly on the simple things of life, we can develop a sense of the forces that act on more distant and less accessible objects. The forces we feel through our fingertips, our fingers, our hands, our arms, our shoulders help us understand the workings of muscles and tendons, of buildings and bridges, of hurricanes and tsunamis, of sink holes and earthquakes, of walls and floors and roofs, of solar systems and universes. The mundane activities of boxing, transporting, and delivering a hot pizza, as well as those of separating, serving, folding, and eating a slice, also provide accessible examples of how problems in strength and stiffness are perceived and how solutions are offered for the design and engineering of complicated systems of all kinds. Indeed, anything made and used serves to illustrate the processes of invention, design, and engineering.

14

Deployable Structures

Tapes Measure

Unlike bicycles, automobiles, locomotives, and other moving mechanical engineering marvels, large civil engineering structures are designed to remain in the place where they were built, as they were built. Thus a bridge becomes a fixed crossing of a river, and a tall building a familiar beacon in a skyline. While the distinction between dynamic and static outcomes is fair, imputing motives to engineers is not, in spite of the witty but cruel distinction that attributes weapons to mechanical engineers and targets to civil engineers. There are simply too many exceptions to the facile distinction for it to be taken seriously, and all of them have to do with the involvement of force.

Bridges are designed to be strong and stiff, but still they move. The Golden Gate Bridge is a good example. Pedestrians crossing it, especially when they are near midspan, can easily feel the bridge bounce as heavy buses and trucks pass. In 1987, the fiftieth anniversary of its opening was declared Pedestrian Day and the roadway was closed to vehicle traffic. Celebrants crowded shoulder to shoulder on virtually every square foot of the structure's roadway and sidewalks, making the resulting mass the largest load the bridge had ever experienced, a condition evidently not anticipated. People on the suspended span felt it sway. The profile of the deck, which under normal circumstances has a distinct camber to it, flattened out noticeably, and engineers observing the phenomenon from a distance expressed genuine concern. Fortunately, the bridge survived its trial by pedestrian, but a seventy-fifth anniversary celebration was not allowed to become another mammoth game of Sardines.

All bridges bounce and sway to some extent. If they could be designed to be perfectly rigid they would be not only prohibitively expensive to build but also unable to accommodate the forces that accompany the expansion and shrinkage of steel and concrete with changing temperatures and loads. Some bridges are deliberately built to move in a gross sense. The bascule span of London's Tower Bridge, which opens to allow the passage of tall ships, is one example. Others include lift bridges that operate like elevators. Even buildings that are the epitome of rigidity sway in heavy winds and earthquakes, and some are designed with parts that move. The 2001 addition to the Milwaukee Art Museum incorporates a brise-soleil consisting of birdlike wings that adjust to shade an atrium from the sun. Such hybrid exceptions to the conventional concept of a building fall into a category known as deployable structures, those that have more than a single appropriate configuration. In Dubai, United Arab Emirates, each floor of a planned eighty-story Dynamic Tower is expected to be capable of rotating 360 degrees and provide changing views to its occupants. At the same time, the tower itself will present a dynamic addition to the skyline when the floors move in a preprogrammed fashion that gives the impression of a drill bit penetrating the sky.

In fact, there are many kinds of deployable structures, and they occur in diverse applications in all branches of engineering. They range from retractable automobile antennas to paper grocery bags to paperboard and corrugated cardboard boxes to tents to folding-wing aircraft to parachutes to observatory telescopes to the gigantic movable roofs that cover sports arenas during inclement weather. Among the most familiar and convenient of deployable structures are umbrellas, many of which can be opened with the push of a button. And when the rain is over, some umbrellas can be restored to a configuration compact enough to be carried in a briefcase or purse.

Many professionals rely on deployable structures. Musicians play bagpipes and accordions. Photographers use expandable reflectors to light their subjects indirectly and collapsible tripods to steady the camera. Carpentry has traditionally been an occupation employing low-tech hand tools, but since carpenters often need to measure something greater in size than their toolbox, the folding rule became a familiar accessory after its invention in the mid-nineteenth century. I have found folding rulers awkward to deploy—with the zig-zag motion requiring a dexterity and coordination I do not seem to possess, at least in combination with speed—and easy to break. The invention and development of the self-straightening, deployable-and-retractable, user-friendly steel measuring tape introduced in the 1920s was a boon to professional and casual carpenters alike. In many instances it has been superseded by laser-based measuring devices, but it maintains its place in the toolbox. It is also a handheld laboratory for demonstrating phenomena relating to force, deformation, and motion.

A U.S. patent for a "coilable rule" was issued in 1939 to Frederick A. Volz, of New Britain, Connecticut, who assigned rights to the tool manufacturer Stanley Works, headquartered in the same city. The Volz rule was operated very much like the retractable tape measure, also made by Stanley, that I keep in my toolbox today. Mine is of modest size—the tape being only twelve feet long when fully extended—but it has many interesting and useful features. My PowerLock model has a sliding plastic button that clamps the deployed tape at a desired length. The color of the blade is "lifeguard yellow," which makes it stand out on a construction site. The end of the tape is fitted with a "Tru-Zero hook," a small piece of steel bent to a right angle. The hook itself may be thought of as a deployable structure within a deployable structure, for it is only loosely attached to the end of the measuring blade by means of a pair of rivets that project through elongated holes. This allows the tip to slide back and forth along the tape a distance equal to the thickness of the hook. When pushed against the inside of a doorjamb, say, the hook comes in contact with the end of the blade, thus making up the otherwise missing initial thirty-second of an inch of the tape markings. When hooked around the end of a board, the hook extends out by that amount, so the inside of the hook serves as the true zero point for the measurement.

My handy tape has seen much use and shows considerable wear. Its blade carries a notice of being protected by a patent issued in 1964 to William G. Brown of New Britain, the rights to which were also assigned to the hometown Stanley Works. The principal focus of Brown's patent is on the plastic film that encases the blade. Though it provides no structural advantage, the film does protect the rule proper. According to Brown's

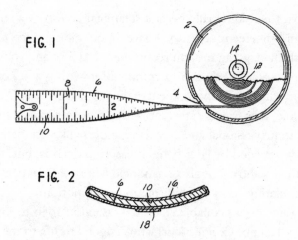

FIG. 1

FIG. 2

Patent drawings for a "coilable metal rule" show (*"Fig. 2"*) how its steel tape is dished longitudinally to give it a stiffness against sagging when in use. *From U.S. Patent No. 3,121,957.*

patent, the surface of uncoated deployable metal rules was subject to corrosion and abrasion, which obliterated the numerals and markings and thus reduced the usefulness of the device. A protective coating of lacquer was often employed, but this only aggravated another important problem with the extensible and retractable tape—namely, it increased the surface friction, which impeded easy deployment and retraction. The friction-lessening coating Brown preferred was made of a "linear polyester film, specifically polyethylene terephthalate, which has proven particularly suitable for this application." My tape measure, which is coated with a DuPont Mylar film, works like a charm, even after decades of use.

The obvious purpose of the spring-loaded self-retracting rule, after that of serving as a measuring device, is to enable its user to deploy and retract it expeditiously. If the rule had to be pushed manually back into its case, the long steel tape would likely bend, buckle, and kink. After repeatedly doing so it might be rendered unusable for future accurate measurement.

The act of extending the tape winds a spring inside the case, storing energy needed to pull the tape back inside. This action provides an opportunity to feel the dynamic effects of an impact force. As anyone who has used such a tape has no doubt experienced, when it is allowed to retract at maximum speed from being extended a good distance, the end hook will slam into the case with considerable force. Not only does this provide a sensible demonstration of the force that a moving object can impart to a stationary one, but also it shows the effect that an off-center force can have. This is easily demonstrated by retracting the tape when the case is on its side on a smooth, flat surface. When the hook hits the case, it will turn it in place, demonstrating in a new context that a linear force applied tangentially can cause a rotational motion. When the case is held in the hand the phenomenon will be felt as a noticeable twist. If the case is not held securely, it can literally spin out of the hand, especially if the tape is a long, heavy one, like the thirty-foot-long, wide-bladed Stanley that I acquired when working on a project of much greater proportions than I had theretofore tackled. The forces felt by the hand holding a steel tape whose end hook slams into its case are dynamic as opposed to static forces, and they demonstrate the magnifying effect of mass combined with velocity. Not a few inventors have patented means to lengthen a tape's life by reducing the effects of repeated hook impact.

An important feature of a deployed steel measuring tape is that it retains a degree of stiffness for some distance from its housing while it remains flexible enough to be manipulated into corners—and, of course, to be retracted into its compact case. These qualities allow both the professional and do-it-yourselfer to measure ceiling height, lumber length, and the dimensions of large spaces without the assistance of a helper. I have watched carpenters and flooring contractors lay down the end of a steel rule at one side of a great room and deploy it while walking along it to the other—and getting an accurate measurement without once leaving a standing position. When finished, they naturally retract the measuring tape into its case—though sometimes so fast it produces a dangerous whipping action—to be returned to a pocket, apron pouch, or toolbox or clipped to a belt.

Much of my deployable rule's behavior is a result of the continuous longitudinal dishing that gives it in Volz's patent terminology "a concavo-convex cross section," which means that it is shaped like a shallow gutter. If it were perfectly flat, it would droop noticeably when extended out far enough. The curvature gives the tape the structural stiffness to resist drooping like a wet noodle. In fact, up to about a foot-and-a-half of extension,

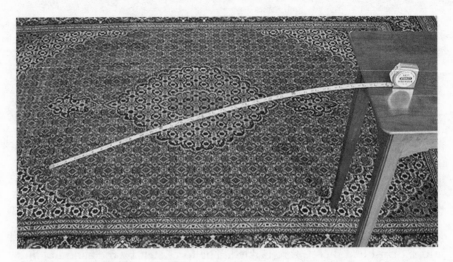

A steel measuring tape extended out to about three feet takes the shape of a classic curve known as the elastica. *Photo by Catherine Petroski.*

its tip deflects only a barely noticeable amount. Deploying the tape further introduces an increasingly obvious curve along the length of the tape— a curve known as the elastica, the analytical determination of which was a subject of great interest among eighteenth-century mathematicians and mechanicians. The elastica of my rule becomes much more pronounced as it is deployed beyond a couple of feet. As it approaches a three-foot extension the tape becomes difficult to hold steady either vertically or horizontally; it flaps and sways at the slightest movement of the casing in my hand. By steadying the casing on a table, I can extend it to about three and a half feet, at which point its tip is about two feet below the tabletop. Extending it farther causes it to buckle and then oscillate like a pendulum for a few cycles before coming to rest in an almost vertical position. If I repeat this experiment with my thirty-foot Stanley held at the edge of my desk, I can cantilever its blade out a distance of about seven feet before it touches the floor thirty inches below.

Structures can also store energy and release it dramatically. This eye-catching behavior enables toylike devices to be used for advertising and entertainment. One such object takes the form of a thin disk about an inch in diameter and shaped like a miniature platter. Although it may not look bimetallic, it is in fact made of two different metals bonded together.

Its concave side is often decorated with the logo of whatever entity gives it out at conventions and job fairs. The disk is "activated" by warming it in the hand and pushing on the convex side until it becomes concave, which is easily accomplished by placing the disk dished-side down across the gap between the index and middle fingers and pressing on the crown with the thumb until it emits a dull pop as resistance to the thumb suddenly drops off. This signals that the energy spent in transforming the configuration of the disk has been transferred to it. When the still warm disk is placed convex-side down on a table, its opposite sides will cool and contract at different rates and the disk will soon snap back into its original configuration. As it does so, the rim will push down quickly on the tabletop, launching the disk into the air. To the uninitiated, it appears to jump spontaneously. The simplicity and mystery of operation of the jumping disk makes it something convention-goers want to keep and so continue to display the logo of its provider, a mechanical equivalent of the internet's pop-up ad.

Another amusing device is the so-called slap bracelet, a preteen fad item of the late 1980s and early 1990s. Made from a nine-inch-long by one-inch-wide piece of spring steel that looks as if it were cut from a retractable tape measure, the bracelet has two stable configurations: the straight one of a deployed tape and the fully circular one that a tape measure assumes when retracted into its case. Slap bracelets were often sheathed in a colorful fabric, which disguised their metallic nature. Just as a deployed measuring tape can be cantilevered out from its case some distance, so the short metal strip will remain straight until forced to change. This will happen if the strip is tapped dished-side down across the wrist, at which time it curls up into a bracelet shape. The forces involved can be gentle enough to feel more like a reassuring tap followed by an encircling caress than a sharp reprimand. Forces are funny like that. How they feel and what they do can all depend upon their human context.

15

Anthropomorphic Models

From Caryatids to Avatars

The dimensions of some ancient royal human body have long been the standard by which we measure all things animate and inanimate. Yet the feet, hands, and fingers of the common folk have served in a pinch as a convenient scale for ad hoc measurements. My own feet are surprisingly close to twelve inches in length. By counting heel-to-toe steps, I can get a very good estimate of the dimensions of a room or field. The width of my thumb at the knuckle is about an inch and so serves to approximate shorter distances. But the history of societies and civilizations is about more than measuring. Architects and engineers used to be one and the same person, sometimes called a master builder, and his creations integrated aesthetics and symbolism with magnitude and structure. The design of temples, monuments, and other sacred structures supplemented primitive measurements with experience, trial and error, intuition, and inspiration.

Consider the proportions of the simple structural and architectural element known as a column. In his *Ten Books on Architecture,* the first-century BC Roman architect-engineer Vitruvius described how the ancient Greeks modeled the Doric column after the average man, the length of whose foot is typically about one-sixth his height. Thus, the diameter of the foot or base of the classical column was established to be one-sixth the height of its shaft, including the capital, which of course represented the head. The most obviously anthropomorphic columns are known as caryatids. They were made in the image of women from the city of Caryae, which sided with the Persians against the Greeks in the Peloponnesian War. Sub-

Caryatids can function architecturally and structurally as columns to support an architrave. *From Vitruvius,* The Ten Books on Architecture.

sequent to the victory of their side, Greek architects designed public buildings with the statue-columns resembling the captured women bearing the load of the structure above them as a reminder of their treasonous acts against Greece. Today, looking at a building that incorporates caryatids into its facade, such as the Erechtheion's Porch of Maidens on the Acropolis in Athens, or the Greek Revival Saint Pancras New Church in London, or the Museum of Science and Industry in Chicago, we can almost feel the weight of the stone above bearing down on the bodies of the women supporting it.

Whatever form in whatever material a column is made—caryatid, Doric, stone, concrete, wood, steel—the weight that it must bear is essentially what rests on its head and presses down through its foot. Engineers call this compression, and it is the principal action that must be resisted also by the blocks of stone in a pyramid, by timber power poles in an electrical grid, and by the steel uprights of a skyscraper. We may not readily see ourselves in the role of those purely functional components, but we can easily

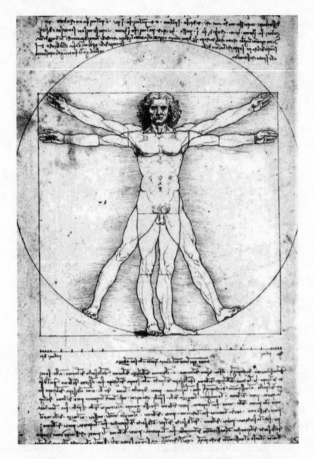

Leonardo da Vinci's drawing of Vitruvian Man shows the proportions and reach of the human body, as based on observations recorded in Vitruvius's *Ten Books on Architecture* as well as on original anatomical measurements made by Leonardo. *Courtesy LeonardoDaVinci.net.*

identify with a caryatid and imagine its pain. We may also feel the forces that tire the arms and legs of Vitruvian Man, whom Leonardo da Vinci described as having "the proportions of the human body according to Vitruvius." To the student of force, this reputedly ideal individual may seem to be shifting his arms and legs because he wishes to take a break from posing spread-eagle to stand at ease. While it may not be literally meaningful to speak of what inanimate materials such as stone and steel feel, engineers do so when they speak of them by using words borrowed from geometry and anatomy. They can imagine themselves as structural stand-ins and anthropomorphize the situation, feeling the forces vicariously.

In a sense, inanimate materials do get tired, especially when they have to resist a force for a very long time or have to repeat an action over and over. And just as elderly people grow shorter with age because calcium leaches from their spinal column, so structural columns shrink in height over time through the phenomenon of creep. Repeatedly loaded and un-loaded beams weaken through fatigue. Engineers use such words, which ordinarily convey human sensations, for the same reason that we use anthropomorphic models of structures to bring them into the sphere of human experience.

The opposite of a compressive force is a tensile one, and it is what we directly experience in the familiar game of tug-of-war. When we engage in such an activity, we can also observe how something without the rigidity of a stone column or steel beam behaves differently when subjected to ten-sion and compression. Before the pair of teams approaches the common rope on which they will tug, it will typically sit sinuous on the ground, like an unkempt garden hose. Once picked up, however, it can be made as straight and stiff as a steel cable, but only as long as it is pulled with sufficient force. Even without an opposing team, we can model our pull and the resistance of the rope by wrapping a piece of string, sewing thread, or dental floss around our fingers and moving our hands apart. As we do, our fingers will feel the floss digging into them, and depending upon its inherent strength and our tolerance for pain, we may or may not be able to break it. But, as William Whewell observed in his treatise on mechan-ics, regardless of how hard we pull horizontally, we will never be able to make it absolutely straight. There will always be a bit of sag.

Not every part of a structure need be in pure tension or compression. Indeed, it is the exception that is. Consider a board thrown across a ditch at a construction site. When workers walk over it, they will feel its flexibil-ity through its bounce, which can be amplified by jumping up and down on it. In bouncing, the board bends, which is a combination of pushing and pulling: the bottom of the board is stretched in tension and the top compressed. If the board were covered with a brittle paint, that on the top would wrinkle and that on the bottom would crack. Anything bent behaves in the same way.

Engineers use the term stress when referring to intensity of force, and it is this intensity rather than the absolute magnitude of a force that determines when something breaks. In a human context, being under tension suggests being stretched or stressed. Individuals who are nervous, easily upset, or under great emotional stress are said to be high-strung. High structural stress, such as that in an overly tightened guitar string, can also result in a breakdown if, as in the guitar, the instrument is strummed too violently. Everything has a breaking point.

Imagine a small bucket suspended from the ceiling by a thin wire, and imagine sand flowing from a hopper into the bucket. Leonardo used such an apparatus, rigged so that when the wire broke the flow of sand stopped. The weight of the bucket plus that of the sand in it was thus the amount of force that broke the wire. Leonardo repeated this experiment on different lengths of the same lot of wire and found that it took a larger amount of sand to break shorter segments than longer ones. This result contradicts classical mechanics, which says that the strength of a wire depends upon its diameter only.

It turns out that Leonardo's conclusion was correct, because wire drawn in his time was imperfect, having a varying diameter and containing random imperfections. The longer the piece of wire tested, the more likely it would have a weak spot at which it would break prematurely. Wire made today is likely to be uniform in its diameter and homogeneous throughout its length and thus should not show great variations in breaking strength. Still, strength tests on wire and other shapes typically do show a spread of strength values because of minute geometrical imperfections, such as nicks and dents, and internal material imperfections that affect the amount of stress that can be tolerated.

John Roebling was a wire manufacturer before he built bridges, but a perceived conflict of interest barred his company from supplying wire for the main cables of his Brooklyn Bridge. He understood that in the later part of the nineteenth century steel wire could be drawn to high standards, but unscrupulous manufacturers often tried to foist off an inferior product. To insure that the cables of the record-setting bridge would contain only good wire, a sample from each reel delivered to the construction site was tested for strength using machines that measured the same thing that Leonardo's apparatus did—the wire's breaking point. When a sample broke at a lower force than expected, the whole lot from which it came was rejected; wire that proved to be sufficiently strong was accepted and the entire reel from which it came was marked for use in the bridge.

Unfortunately, the supplier devised a scheme whereby an accepted reel of wire was replaced by a rejected one as it was being moved to the construction site. This subterfuge was not discovered until a good deal of poor wire was incorporated into the cables. Rather than disassemble them to test and remove inferior wire, Washington Roebling, who had taken over as engineer-in-chief when his father died, decided to use more certified good wire than originally planned. To this day, the Brooklyn Bridge contains inferior steel, but since each individual wire carries a portion of the total force in the cable, if a weak one breaks there are several stronger ones nearby to take up the slack. In other words, there is strength in numbers, even if some of those numbers are weak.

A single force can produce a small or large stress depending on whether it is spread out across a surface or concentrated on a small part of it. When our cat Ted was a young kitten, he would wake us in the morning by jumping around on the bed. He seemed to be playing out some instinct to pounce on prey, and when his little front paws came down on one of my arms or legs, the discomfort I felt was surprisingly great for such a small animal; it was a demonstration of the magnifying effect of a dynamic force concentrated over a small area. No wonder such an action stuns an animal's prey into submission. It doesn't take a cat to demonstrate the effect of concentrating a force on a small area. We can do so by just pushing a thumb and forefinger against each other in a pinching action. Regardless of exactly how the fingers meet, the force will be about the same, because it is produced by the same muscles. Most people do not have strong enough finger muscles to push so hard that they feel pain. However, if instead of using the ball of the thumb, we press the thumbnail against the flesh of the forefinger, we will feel a much more concentrated sensation that might, in the case of a pointed fingernail, be said to approach pain. We stop pressing because our forefinger begins to hurt and people generally have an instinctual aversion to harming themselves.

Pain and pressure have been described as "among the last frontiers of scientists' efforts to describe the molecular basis for sensations." Whereas our senses of smell and taste are localized in the nose and tongue, thereby providing a hint as to where to focus research, pain and touch can be perceived throughout the body. This made the discovery of "key mechanisms of how people sense heat, cold, touch and their own bodily movements" worthy of a Nobel Prize. Indeed, in 2021 the prize in physiology or medicine was awarded jointly to the physiologist David Julius and the molecular biologist Ardem Patapoutian, who independently achieved a breakthrough in understanding the phenomena.

Many an engineered structure, though it may not be sentient, involves the interaction of a lot more complicated forces than those of a couple of fingers pressed together or a simple column, length of rope, or board across a ditch being compressed, tensed, or bent. To get a feel for the forces and how they interact in such a structure, analogs and models more elaborate than fingers are necessary.

Dorton Arena, located on the state fairgrounds in Raleigh, North Carolina, is an unusual structure. Completed in 1953, it consists of a pair of reinforced concrete archlike members, each inclined at a shallow angle to the ground and crossed like the legs of a camp stool. Stretched between the tops of them is a network of steel cables that supports a lightweight roof. The weight of the backward-leaning arches keeps the cables in tension, and the tension in the cables in turn keeps the arches from falling over backward. The arrangement is not unlike opposing teams engaged in a tug-of-war. The innovative building structure was designed to provide a column-free interior space suitable for horse and livestock shows, hence its having been referred to early on as a "cow palace." The structural principle it embodies has been compared to that at play in a foldable director's chair, in which the canvas seat is slung between two sets of legs that cross at their midpoint. People sitting in such a chair are supported by the canvas seat the way they would be in a hammock. Their body's weight induces a tension force in the canvas, and that force pulls inward on the tops of the legs. At the same time, because the legs are joined together loosely where they cross, they act like a pair of scissors and their bottoms are pushed outward. The combined pull-and-push action makes the chair work.

The engineer William Baker has used an anthropomorphic model to illustrate how his structural design for Burj Khalifa resists the forces of the wind. When constructed as the tallest building in the world, the Dubai tower represented a new paradigm in super-tall building design. Since it was not simply a scaled-up version of previous record holders, such as the boxy Sears Tower or tiered Taipei 101, Baker found it helpful to have a readily understood model of the Burj to explain its ability to resist high wind forces. He found a ready analog on the streets of Chicago, the city in which his Skidmore, Owings & Merrill office is located: a man with an open umbrella held against the horizontally blowing wind and rain steadies himself against the onslaught by extending one leg behind his body to serve as a brace or buttress. Indeed, Baker describes the structure of Burj Khalifa as a buttressed core—meaning that the tower's central shaft containing the building's elevators and their attendant lobbies is braced by the tapering stacks of floors that radiate out from it in three directions. The model is an excellent vehicle for giving people, most of whom will have wielded an open umbrella against windblown rain, a feel for how the tower stands up against forces that otherwise might bring it down.

Human analog models can also give us a feel for what went wrong when a structure fails. The Kansas City Hyatt Regency Hotel was designed with guest rooms located in a tower on one side of a lobby atrium and meeting and assembly rooms in a separate block on the other side. Elevated walkways spanning the atrium not only kept much of the back-and-forth meeting traffic off the lobby floor but also provided architectural features to break up an otherwise empty vertical space. Two of the walkways were suspended one above the other and held in place by nut-and-washer assemblies on threaded steel rods anchored in the roof. For reasons of convenience during construction, each long rod of the original design was replaced by a pair of shorter ones. The upper walkway was still supported from the roof, but now the lower one was supported from the upper. In the summer of 1981, the newly opened hotel became infamous when without warning the pair of heavy steel walkways fell to the floor, which was full of dancers and revelers. The toll of 113 people dead and many more injured made it the worst structural accident in American history. In the case of the Hyatt Regency walkways, a simple anthropomorphic model helps explain what happened and why. Imagine each walkway as a person

A marine pausing during a rope-climbing exercise
would be able to hang on only as long as he could
maintain his grip on the rope. If another marine was following him up the rope and suddenly changed his grip
from the rope to the ankles of the higher marine, the
grip of that marine would have to double in strength.
If he could not hold on, both of them would fall to the
ground. *Library of Congress, Prints & Photographs
Division, FSA/OWI Collection, LC-USE6-D-005795.*

and the rod as a rope. The original design is represented by two men holding onto a single rope hanging from the ceiling of a gym. The single rope
is strong enough to support the weight of both of them, but each man
must hold on with a grip sufficiently strong to keep himself in place. Now
imagine if the lower man were to shift his grip from the rope to the legs
of the person above him, which would be analogous to the change made

in the support system of the walkways. The change makes no difference to the rope, but it means the upper man now has to exert a grip strong enough to hold in place not only himself but also the man hanging onto his legs. If the top man's grip cannot support the added weight, both people will fall to the floor. In the hotel, a fourth-floor beam held in place on the support rod by a nut-and-washer connection could not continue to support the weight of both walkways plus the unusually large crowd using them, and so beam separated from rod, just as the hand of the overburdened rope holder could not hold its grip. Once this happened at one structural connection on the upper walkway, others had to take up the slack, which they could not do, and a progressive collapse ensued.

Being able to feel the forces acting in an unusual structure, even if only by proxy, gives engineers, students, and laypersons alike a visceral understanding of its workings. If we can actually be part of the model and feel the forces directly, then so much the better. Probably the most iconic anthropomorphic model of a major engineering work is one associated with the first bridge to span the wide estuary of the Forth River near Edinburgh, Scotland. As the North British Railway pushed farther northward, fixed crossings were needed for the firths (estuaries) of the Forth and Tay Rivers. Without such crossings, the need to transfer rolling stock and passengers to and from ferries at each riverbank slowed progress up the east coast. The railway engineer Thomas Bouch was commissioned to design bridges to span the firths, and construction of the one across the shallower Firth of Tay was completed in 1878. It was not a particularly daring design, and the bridge was remarkable mainly for its length of almost two miles. Unfortunately, the longest and highest girders of the bridge collapsed during a storm in December 1879, killing seventy-five people on a train making the crossing. What today might be called a political cartoon showed an anthropomorphized "Spirit of the Storm" pushing and punching the bridge into submission. The structure might be said to have felt the force as the spirit moved it.

A court of inquiry found that the bridge was "badly designed, badly constructed and badly maintained," and its engineer was discredited. Construction was halted on the suspension bridge that Bouch had designed to cross the deeper Firth of Forth, and subsequently an entirely new design for that crossing was commissioned from the firm of the distinguished engineer Sir John Fowler and his younger partner Benjamin Baker, who himself would be knighted upon the completion of the project. Baker, who years earlier had published a book on long-span railway bridges and was the lead designer of the new bridge, confessed to the enormity of the task: "If I were to pretend that the designing and building of the Forth Bridge was not a source of present and future anxiety to all concerned, no engineer of experience would believe me. Where no precedent exists, the successful engineer is he who makes the fewest mistakes." Fortunately for engineers, it is also precisely where no precedent exists that they tend to be the most careful and hence successful.

Baker's design was based on the cantilever principle, which Galileo had articulated two and a half centuries earlier but had not been very much used for bridges in Britain. But Victorian engineers were open to new ideas and the public was quite enamored of public lectures that explained the latest developments in science and engineering. In 1850, when the innovative Crystal Palace was under construction in London's Hyde Park, an on-site lecture with props explicating the design's structural principles— as well as those of some unusual contemporary bridges—was presented before a lay audience. Benjamin Baker would speak on the Forth Bridge at the Royal Institution, the same venue where in 1859 Michael Faraday had given his famous Christmas lectures on the forces of matter.

In preparing his lecture, Baker "had to consider how best to make a general audience appreciate the true nature and direction of the stresses on the Forth Bridge, and after consultation with some engineers on the spot, a living model was arranged." The model, which represented one of the spans of the bridge and came to be known as the "human cantilever," would prove to be the most memorable part of the lecture. Although Baker did not claim the idea to be wholly his, the famous model has been closely associated with him ever since.

The tableau vivant employed three men, a pair of chairs, four wooden struts, two pallets of bricks, a swing seat, and some rope to connect the latter components to each other. Two of the men sat upright in the chairs and grasped the tops of a pair of struts, the bottoms of which were notched to fit against the edge of the chair seat. The triangular arrangement so formed on either side of each man represented the portion of the bridge structure symmetrically cantilevered out from the bridge towers. To the

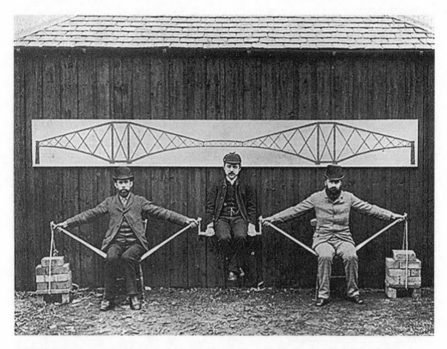

A "human cantilever" was employed by the engineer Benjamin Baker to illustrate his 1887 lecture explaining the novel structural design of the record-setting steel cantilever railroad bridge then under construction across the Firth of Forth near Edinburgh, Scotland. *From* Engineering News. *Photo by Evelyn George Carey.*

tops of the inside struts were hung the sides of the swing seat, which represented the suspended central portion of a bridge span. Each outside cantilever, consisting of the man's other arm and associated strut, was tied to a pallet of bricks that provided the counterweight to balance half the combined weight of the suspended seat and the man who occupied it.

In Baker's lecture, a large drawing of the bridge span was displayed behind the human model, thereby making it self-evident what parts of the model corresponded to what parts of the bridge. In Baker's own words, "When stresses are brought on this system by a load on the central girder, the men's arms and the anchorage ropes come into tension and the sticks and chair legs into compression. In the Forth Bridge you have to imagine the chairs placed a third of a mile apart and the men's heads to be 300 ft. above the ground. Their arms are represented by huge steel lattice members, and the sticks or props by steel tubes 12 ft. in diameter and 1¼ in. thick."

Ironically, the "living model" was not demonstrated live at the lecture but rather was projected from a lantern slide, which became the source for the famous illustration of it. The American bridge engineer Thomas C. Clarke, who visited the construction site in 1887, was acknowledged as providing the photograph from which the American trade journal *Engineering News* made an engraving to accompany its report on the lecture, which was described as being "received with loud and general applause." This appeared within about three weeks of the lecture and about a month before it was published in the British publication *Engineering*, showing the speed with which engineering news, knowledge, and documents could travel across the Atlantic in the late nineteenth century. In the twenty-first, it remains commonplace for engineers and laypersons alike to visit the bridge and replicate the famous living model. When I visited the site in 2003, there on the grounds of the visitor center were a pair of steel chairs and all the rest of the apparatus needed for a trio of bridge buffs to participate in reanimating the model while looking out at the structure itself.

The apparatus was subsequently relocated across the firth to South Queensferry, where it sat in the back court of the Orocco Pier restaurant. According to Roland Paxton, engineer, scholar of engineering, bibliophile, and lover of bridges generally, the replica was "on loan to the restaurant from the town's Forth Bridge Memorial Committee to whom it was donated by the Forth Bridges Visitor Centre Trust in 2012." The final meeting of the trust was commemorated in a photograph of the trustees posing with a re-creation of the human model, with Chairman Paxton sitting in the catbird seat, albeit not nearly as high off the ground as the Japanese exchange engineer Kaichi Watanabe did in the original model.

When it comes to feeling forces, there are some other interesting differences in the two demonstrations posed 125 years apart. In the 1887 tableau, the human participants representing the towers are sitting with their legs close together, no doubt in part because the struts were confining their thighs. In the 2012 re-creation, the gentlemen in the chairs have their legs spread apart, evidently because the struts were not so confining. In the nineteenth-century model, the chair-sitters are grasping the struts with their palms facing the camera—and with what appear to be firm grips—suggesting that their arms are indeed in tension, as they would be expected to be to hold the struts at the angle necessary to support the

Trustees of the Forth Bridges Visitor Centre Trust re-created the human cantilever in 2012. *Prof. Roland Paxton, Chairman, Forth Bridges Visitor Centre Trust (1997–2012).*

swing seat as high above the ground as it is. In the twenty-first-century re-creation, however, the gentlemen in the chairs are grasping the struts with their palms facing away from the camera, a less natural and efficient way to provide the forces an observer might think necessary to hold up the struts, which are less steeply inclined.

Furthermore, in contrast to the original living model, in which the chair occupants grasp the struts as close to their far end as possible, thereby exerting the minimum force needed to form the isosceles-triangle cantilevers, the trust members are holding the struts at about midpoint, which does not give the model the verisimilitude it should have. In addition, the way they are holding the struts suggests that rather than supporting them, the seated men are actually leaning on them, thus putting their arms not in tension but in compression. This is especially evident in the open right hand of the gentleman in the left chair: he does not even have his fingers wrapped around the strut, and he appears to be just resting on or pushing down on it with the heel of his hand. As for the gentleman in the chair on the right, his arms appear to be crooked a bit, further suggesting that they are not being fully extended in tension. The men are posing with the model apparatus but do not appear to be experiencing or conveying a sense of the forces involved to the extent that the original model was designed to do.

In the original model, Watanabe holds his hands close to his side, grasping the swing seat, maybe to secure his balance. In the 2012 re-creation, however, Paxton is resting his hands on the struts, perhaps because the other gentlemen have left room for him to do so, but the potential effect is to introduce confusion among viewers of this reenactment of how the real structure works. In fact, given that the gentlemen in the chairs do not appear to be holding the struts with any tension in their arms, the actual structural action of the center part of the model appears to be more that of an angled arch or A-frame than of a cantilever. What holds the swing seat and its living load in place is the chairs and their occupants serving as abutments rather than as towers.

Finally, in the Baker demonstration the notched lower end of each strut rested on a side edge of the chair seat, forcing the seated gentlemen to keep their legs close together. However, in the visitor center trustees model this does not seem to be the case. Extrapolating the visible portion of any one of the struts suggests that its lower end does not rest on the edge of the chair seat but rather intersects the chair structure somewhere under the seat, thereby giving more sidewise leg room to the participants. Closer inspection of the juncture of the upper end of a strut with the suspended seat shows it to look like a welded joint. This indicates that the struts are metal tubes and suggests that such connections may have been employed at least in part to keep the apparatus complete and so always at the ready to be used by visitors to re-create for a photograph rather than to provide the opportunity to feel the forces at play. Engineering and a full understanding of it are in the details.

The success of the Forth Bridge as an engineering project and as a structural phenomenon popularized to a great extent by the human model of it made the cantilever form appeal to engineers and nonengineers alike. Wherever a new long-span bridge was under consideration in the late nineteenth and early twentieth centuries, especially if it was to carry railroad trains, the cantilever was seen as a viable if not the preferred alternative to the suspension bridge. Cantilever designs were proposed to span the broad expanses of the Hudson River at New York City and the harbor at Sydney, Australia.

But not everyone was totally enamored of the Forth Bridge. The distinguished American railway engineer Theodore Cooper thought it to be "the clumsiest structure ever designed by man; the most awkward piece of engineering" ever to have been constructed. He considered it grossly overdesigned and believed that an equivalent crossing of the firth could have been built for about half the price. When he became involved as principal consulting engineer for a project to bridge the Saint Lawrence River near Quebec City, he saw his chance to prove his claim. The original plan called for a cantilever with a main span of 1,600 feet, but Cooper argued successfully for an altered design with piers closer to the riverbank, thereby making them less likely to be struck by ice floes and reducing the cost of construction. This increased the main span to 1,800 feet, which incidentally would make the Quebec the longest-spanning bridge of any kind anywhere in the world—and a fitting magnum opus to the aging Cooper's career.

There were many similarities between the designs for the Forth and Quebec Bridges, but in keeping with Cooper's feelings about the Forth being overdesigned, the Quebec was to be a much lighter structure both aesthetically and structurally. By August 1907, the south cantilever had been built out to a length of about 735 feet. Unfortunately, before it could be extended any farther, it collapsed into a heap of steel, killing seventy-five construction workers.

The failure was attributed to several factors, including a young engineer who seems not to have had a feel for the forces he was calculating. He underestimated the total weight of the bridge and consequently specified in his plans steel members that as the bridge was being built would be overstressed and insufficiently stiff. Cooper had not provided proper oversight of the project to catch these errors in time to save the bridge. About a century after the collapse, the Swedish consulting engineer Björn Åkesson, in his book *Understanding Bridge Collapses,* provided a detailed analysis of the failure by employing a model that did not involve the hardware of chairs, struts, ropes, and bricks but used a digital stick structure to represent the cantilevered span and a virtual giant as a counterbalancing avatar. He used computer software of the kind that comes with Power-

A photograph of the Quebec Bridge shows it under construction in 1907. The weight of the span cantilevered out above the Saint Lawrence River was balanced by that of the landward portion, with the steel links connected to the top of the tower being in tension and the struts at the bottom in compression. If these components were not strong enough to resist being pulled apart or being crushed or bowed by the growing weight of the lengthening span, a collapse would occur, as in fact it did not long after the photo was taken. *Credit: Canada. Patent and Copyright Office, Library and Archives Canada. British Library, Digitised Manuscripts, HS85/10/18815.*

Point and similar presentation programs to draw a schematic of a cantilever span under construction balanced by the weight and strength of the avatar representing the equilibrating forces provided by the landward span. The giant's anatomy consisted of eight ellipses, representing torso, head, and limbs articulated as in those of an artist's wooden mannequin. The hands grasped the top of the cantilever, the arms pulled back, and the body positioned itself to maintain balance, while the feet and legs pushed against its bottom to maintain position. The model showed how the cantilever was kept in place by this action, but it did not account for the fact that as more steel was added to the truss to extend it farther out across the river, the virtual giant had to grow in strength and weight or redistribute it to keep the cantilever horizontal.

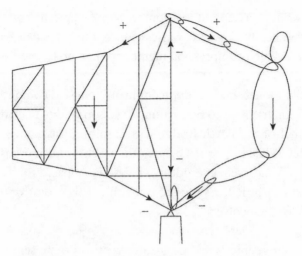

The forces involved in balancing the cantilevered portion of the Quebec Bridge may be felt vicariously through this humanoid model. The avatar will hold the structure in place only as long as its muscles, tendons, and bones can exert and resist the tension forces pulling on its hands and arms and the compression forces pushing against its feet and legs. If they could not, the bridge would collapse, as it did on August 29, 1907. *After Åkesson,* Understanding Bridge Collapses.

If we identify with the virtual giant and its stance, we can feel the kinds of forces involved and see why the bridge failed. As the cantilever grows in size, our arms and legs have to provide correspondingly larger forces of tension and compression, respectively, to maintain equilibrium. This means that our hands must grasp more firmly, our arms resist an increasingly greater pull, and our feet and legs lock in place and push harder and harder as more and more steel is added—and they must do so until the cantilever meets its counterpart extending out from the other side of the river and leans against it. Only then can we relax and let go. Even a virtual giant should have a limit to its strength and endurance, however, so taking too much time to build the bridge may doom it. How it might fail would depend on where in its body the giant lost its strength. The hands or arms giving out would correspond to a tensile failure (such as losing a hand grip or arm strength, perhaps representing a fracture), and the legs giving out would correspond to a compression failure, such as a foot being crushed or a knee buckling. It was essentially the latter that happened to the real bridge when compression members could no longer bear the load placed on them.

Humanoid models may seem quaint in an advanced computer age, but virtual ones might prove to be a natural. Åkesson credits Benjamin Baker's human model of the Forth Bridge with inspiring the giant's cantilever-balancing act. While this virtual behemoth, which Åkesson terms an "imaginary 'being,'" may seem as crude as a rag doll compared to the flesh-and-bones Victorians in Baker's model, it is sufficiently evocative to convey a feel for the forces involved.

Just as the visitor center's trustees inadvertently misrepresented the physical Forth model, so engineers can err in applying sophisticated computer models. In fact, all models must be constructed, used, and interpreted with care, lest they mislead and provide a false sense of how a structure works and what the forces within it are. The Forth Bridges visitor center outdoor model apparatus was no doubt developed to be durable and user-friendly. By making its components out of steel, the designers ensured that there was little chance of them breaking or wearing out. With the chairs bolted to the concrete base, counterweights fixed to it, and the lighter components hinged to the heavier, the parts were unlikely to become separated, lost, or stolen. The model must have seemed to be perfect for its purpose. It was also very user-friendly, in that all a trio had to do was to sit in the chairs and on the swing seat and have its picture taken. The hand-arm coordination required of the participants in the original model was not of the casual tourist.

However, as the photograph of the trustees reenacting the model shows, the seemingly user-friendly model was also conducive to misuse, even by experienced structural engineers. All the men had to do was assume their places on the interconnected seats so their weight alone could lock the center swing in place. In contrast, the original separate pieces of apparatus would have required the men to take care in setting up the model and holding their pose so they could truly experience the forces involved. When the model evolved into a mere prop for a photo op, it lost its efficacy. It was too easy to sit in the chairs and on the swing seat without worrying exactly where or how the hands and arms held the struts or felt the resulting forces. Such unthinking errors might be even more easily committed and consequential when using computer models not just as demonstration aids but as calculational tools.

16
Visible and Invisible Hands

Wind, Warp, and Woe

In his *Wealth of Nations,* the Scottish economist Adam Smith wrote of an invisible hand that produced an unintended though not necessarily bad result. Even if unseen and unknown, such a hand could leverage the self-interest of an individual consumer into a benefit to a nation's entire economy. Invisible hands work levers in all sectors of society, including those explicitly involving physical structures. We sense the presence of an invisible force whenever the wind blows. It unfurls limp flags into proud banners and scatters falling leaves as if it were sowing fresh seed. It also can tear a flag to shreds and render a seeded field fallow. To avoid being blown away ourselves by a strong and steady wind, we bend our body into it, as if weighted down by a burden on our back. Unfortunately, if we are not prepared to respond to changing conditions, we might fall on our face when the wind itself stops to take a breath.

"Who has seen the wind?" the English poet Christina Rossetti asked, and answered, "Neither you nor I: / But when the trees bow down their heads, / The wind is passing by." When we exhale our warm breath on a cold day, it becomes visible, as do the air currents in the smoke from a candle, cigarette, or chimney. A helium-filled balloon also attests to the forces of which the air and wind are capable. The reason it floats upward is, of course, because the inert gas is less dense than air. The heavier air that is displaced by the lighter balloon imparts a buoyant force to the balloon. It wants to rise the way bubbles do in a glass of champagne. Warm air is also less dense than cooler air, and so a balloon filled with hot air will rise.

Leonardo had thought about the forces causing hot air to rise, and so did the French brothers Montgolfier in the late eighteenth century. In observing smoke going up a chimney and clouds floating in the sky, they saw the possibility of a hot air balloon. Their first, which was launched in 1783, measured about forty feet in diameter. A bunch of smaller balloons can accomplish the same end. If we tie enough party balloons to an aluminum lawn chair, say, and do not hold it down, it will float away. If there are a very large number of balloons, the chair can rise even with someone sitting in it. This was demonstrated in 1982 when a California man tied more than forty eight-foot-diameter weather balloons to his chair and rose to an altitude of about fifteen thousand feet, where the wind took over. He inadvertently dropped the BB gun that he had carried with him to shoot out balloons when he wanted to descend and so floated for hours, crossing into controlled airspace and generally creating a dangerous situation for himself, airplanes in the area, and people on the ground. As do all balloons, his gradually lost their charge and so their buoyancy, and the adventurer wound up tangled in some power lines, from which he was extricated and promptly arrested.

Airplanes take off into the wind, and it is the shape of their wings that exploits it and enables them to fly. Two-by-fours can be emboldened by hurricane winds to smash through windows and break down doors as if they were battering rams aimed at a castle gate. Even toothpicks on a table of hors d'oeuvres can be taken up and flung like spears at exposed olives and flesh alike. Buoyed by social winds, a person can get involved in movements that may appear to be going in the right direction but, like Dorothy caught up in a tornado, find themselves ending up far from Kansas. Once back home, the hapless traveler may find his or her town greatly transformed. Physical structures that should have weathered the storm the way corn in the fields did may be altered beyond recognition.

The destructive force of the wind was long known, but how great its push could be was not always properly quantified. Gustave Eiffel, who was basically an engineer of bridges, understood how important it was for his Paris tower to have the strength to resist being toppled over by the wind, and that guided its very shape. Not all engineers were so prescient, however, and their structures succumbed. The collapse of the Tay Bridge in 1879, as we saw in the previous chapter, was attributed to the irresistible

In the wake of the 1879 collapse of the Tay Bridge at Dundee, Scotland, the force of the wind believed to have brought it down was personified as "The Spirit of the Storm," shown here pummeling the wrought- and cast-iron structure. *From* The Illustrated London News.

force of a storm anthropomorphized as a giant muscleman pushing it over. One popular press illustration of the event depicting the brute pummeling the structure was captioned, "Veni, vidi, vici." The effects of gusting winds may be likened to the waves of attacks from an advancing army. They can destroy everything in their path and leave a battlefield littered with lifeless defenders. So it can be when the winds of change blow too strongly.

A more realistic forensic analysis of the Tay Bridge failure attributed it to the inability of the structure to withstand the forces imposed on it. A rational inquiry placed the blame squarely on the shoulders of its chief engineer, who should have been able to see the big picture, including the invisible forces lurking in the background. Every social structure, whether corporation, institution, or government, has its own chief engineer, albeit known by another name. Regardless of the nomenclature, it is the chief executive officer with whom the buck stops. Societal, commercial, and cultural revolutions are as engineered as bridges, and as such they too can be badly conceived, badly formed, and badly run—and unable to withstand a hostile takeover. A confluence of forces can leave behind the debris of unintended—or insidiously intended—consequences.

Almost two centuries to the year after the publication of *The Wealth of Nations,* the business historian Alfred Chandler published *The Visible Hand,* in which he described a managerial revolution in the United States. In his view, corporations were much more than their CEOs. It was the whole system of management that very explicitly controlled the forces of production from raw materials to finished artifact and beyond, as exemplified in the operation of such companies as Ford, General Electric, and Standard Oil. Forces may be invisible or visible, physical or metaphorical, but their effects are always real.

Near Da Nang, Vietnam, there is a graceful gold-colored pedestrian bridge set among the natural beauty of a piedmont. Symbolically, Golden Bridge is suggestive of a bar of bullion being presented to ordinary people by a giant god who resides within the rough terrain. In fact, the hands of the god that appear to emerge from the underlying rock are made of wire mesh and fiberglass finished to look like stone. The bridge was deliberately designed to give visitors stunning views of the area. But what can be inferred about the giant's full anatomy and purpose from its hands alone?

The situation evokes the allegory of the blind men and an elephant, a tale said to have originated among cultures of the Indian subcontinent. In one version, each of six sightless men touches a different part—but only that part—of the elephant and so forms a very limited idea of what the whole animal is like. The man who explores the trunk imagines it to be like a large snake; the one who holds the tail thinks it ropelike; the man who touches the body thinks it a solid wall; a tusk reminds the man touching it of a spear; the ear is like a fan; a leg a tree trunk. When they compare notes and produce a composite picture of their explorations, they come up with a strange animal indeed. The conclusion that we cannot know the whole by experiencing only a single part of it is incontrovertible.

But what if six blind people were asked to characterize an inanimate object in the same way, relying only on their sense of touch to grasp and prod a single part to feel its shape, texture, and stiffness? What would they

The Golden Bridge, located near Da Nang, Vietnam, appears to be supported by the hands of a giant who is said to reside within the granite hills. *Hien Phung Thu/ Shutterstock.com.*

make of, for example, an old-time canister vacuum cleaner, the kind with a long cylindrical body that could be dragged around by its hose?

If the machine is operating, a hand on the hose would feel something flowing through a flexible ribbed tube, suggestive perhaps of a small elephant taking a long, deep breath through its trunk. A person holding a hand against the tube's smooth end would feel suction, perhaps suggesting the snout of a scavenger or a pollinator. Depending on what if any cleaning tool was attached, it might feel like the mustache of a walrus or the snout of an anteater. A hand against the other end of the object would feel air rushing out, maybe suggesting the blowhole of a very large mammal. Someone grasping the electrical cord would feel a thin, smooth, flexible, and somewhat wriggly thing that could bring to mind a long earthworm; the oddly shaped plug at its end could be a skull, its prongs teeth—or a paw with long claws. A hand on the vibrating canister body might sense it to be metallically cool on one end, warm on the other, perhaps where its heart was beating rapidly inside an armored exterior, hard

What might each of a group of blind people touching only one part of a vintage vacuum cleaner imagine the whole to be like? *iStock/OlegPhotoR.*

and unyielding like the shell of a tortoise; a hand lifting the entire appli-
ance would experience a heavy and unwieldy body, perhaps suggestive of
some kind of slippery fish throwing its weight around. What kind of drawing of the whole might the committee come up with?

What we feel when we touch something depends in part upon the condition of our hand and what we do with it. Unless we have magnets embedded in our fingertips, we are not likely to be able to tell the difference
between steel and aluminum. We may be able to sense texture, which can
be imitated in a variety of materials, but we should be able to tell a metal
from a nonmetal because the former will generally be cooler to the touch.
If we can knock on the surface of the thing, the sound we hear should be
helpful. We can tell a lot about the springiness or stiffness of an object by
the way it yields under our fingers or palm—or the way it droops, though
it may take two hands to determine that. What parts we feel may or may
not include the object's most distinctive features. We cannot know what
we do not sense.

No matter how simple or complicated a force in isolation or in combination may be, it is typically exerted on an object in one of two basic ways.
The first is by direct contact, as when our fingers manipulate a shirt button through a buttonhole or when an open hand pushes on the back of a
car stuck in the snow. The second way is by action at a distance, as when
gravity pulls us back down to earth when we leap for joy or when a magnet causes a straight pin in its range to jump up off a table. Contact forces

are tangible; gravitation and magnetism have been said to possess an "occult quality," in that the force is transmitted from one object to another through thin air. There is one kind of force that may seem to fall somewhere between the two, in that it is effected by contact and it is felt to be pushing against us, but the body doing the pushing is not readily visible. This is the force of the air itself when it manifests itself through the pressure of the wind.

We can experience the force of the wind when we stick our hand out the window of a car traveling down the highway. At high speeds, we can feel the air pushing the open hand at the end of our arm toward the rear of the car in a way akin to a rock pulling the end of a cantilever beam down toward the ground. But unlike the rock, which is dead weight on the beam anchored in a stationary wall, the open hand is moving through the air, which tends to flap it around. We can hold our hand in a more stable position by facing our palm down to make the hand more a blade than a sail. If we point our fingers into the wind and let our hand flex up and down like a dolphin pacing a boat, we will feel our hand alternately rising and falling in the dry ocean of invisible but sensible wind. The phenomenon whereby a hand flips from one configuration to another, like an umbrella suddenly turned inside out in a fierce storm, is known to engineers as instability. The spontaneous movement of a structure from a desired configuration to a less desirable one is the manifestation of an instability; it is something that engineers generally want to avoid, because it can happen without warning and leave the structure in a nonfunctional or damaged or failed state. There are always exceptions, of course: the bimetallic jumping disk works precisely because it is activated by being deliberately buckled and being unstable in that configuration.

As late as the mid-nineteenth century, the magnitude of the force exerted by the wind was a matter of some confusion, debate, and downright ignorance, as evidenced by the collapse of the Tay Bridge. Years earlier, Thomas Bouch, the engineer of that bridge, had sought authoritative advice on the nature of the wind force at the Firth of Forth, where he was designing a second bridge for the North British Railway. Astronomer Royal George Airy advised him that an average wind pressure of no more than about ten pounds per square foot of area should be expected, but with gusts producing localized pressures of forty pounds per square foot. Bouch evidently used the lower number for his calculations at the more northerly Tay site, which proved to be a gross underestimate of the wind pressure there. As we have seen, after the Tay Bridge collapsed, Benjamin Baker was given the task of designing a new Forth Bridge from scratch precisely because the train-riding public (and the railroad itself) had lost confidence in Bouch's engineering ability and judgment. Baker designed the new bridge the way he did precisely because of concerns over the wind. For pressures, he did not rely on expert opinion but rather developed an apparatus to measure the actual pressure of the wind on a flat surface set up near the bridge site. The apparatus was analogous to a hand held out the window of a moving vehicle; the force was quantified by measuring how much springs backing up the flat surface were compressed. As the Wright brothers understood, the effect of the wind is the same whether it blows against a stationary object or the object moves through still air with the wind's velocity, and they exploited this knowledge in constructing a crude wind tunnel in their bicycle shop to study the interplay between the wind and the shape of parts of their flying machine even before it flew.

To make the Forth Bridge not only be but look capable of standing up against any blustery giant, it was designed to bear almost no resemblance to the failed Tay. Whereas the superstructure of the ill-fated bridge rose strictly vertically, that of the Forth has battered side columns, meaning that they slant inward as they rise. Everyone knows not to stand straight upright in a windstorm; we instinctively spread our legs to brace ourselves. Such a stance was known in portraiture as a Holbein straddle, after the artist to the court of Henry VIII. It was Hans Holbein who painted full-length portraits of the monarch and others standing with their legs spread apart. Giving the Forth Bridge a straddle conveyed visually a sense of strength and stability.

Some of my most memorable encounters with the wind occurred in Chicago, especially when I was walking on streets running perpendicular to the shore of Lake Michigan. In the Loop area, the tall buildings that line these streets make them veritable canyons through which the wind rushes with great force. Walking toward the lake, I have been practically stopped in my tracks by gusts of wind that checked the momentum of my body. When the wind blew steadily, I had to tilt my body forward and hold my torso at an angle to the sidewalk in order to balance the moment of the backward force of the wind with that of the downward force of my weight. When a steady wind suddenly let up, quick reflexes were required. At intersections, the wind, made visible by the dust and newspapers it kicked up, curled around behind my back, rocking me on my feet. I had to brace myself from gusts, and I did so against sideways ones by assuming a Holbein straddle. Walking in such conditions removes any doubt about the real and tangible force that can appear out of thin air.

In 1887, when the Forth Bridge was under construction, Benjamin Baker wrote to William Unwin, professor of civil and mechanical engineering at the City and Guilds College in London, where strength-testing equipment was better than that available at the construction site. Baker let Unwin know that he was sending specimens from a broken steel bar so that Unwin could determine if the material's properties varied sufficiently to explain the failure. In his letter, Baker expressed incidentally the wish that Unwin could be on the bridge "some windy day to note how erratic the pressure is." According to Baker, all the workmen atop one of the 370-foot-tall towers were "imprisoned" one night "owing to wind pressure rendering it impossible to get down either by cage or stairs. It was a funny place to sleep." Baker also wrote of observing during a gale how an aneroid barometer that he was carrying about on the structure gave different readings at different locations on the girders.

Contemporaneously with the design and construction of the Forth Bridge, Gustave Eiffel was also struggling to understand the full nature of wind forces. His company had been specializing in designing and constructing iron-arch bridges to carry railroads over deep river valleys. As he wrote in a memoir, engineers had to use safety factors that "had no scientific basis." Taking into account the force of the wind on tall structures was especially problematic, for there were many open questions, among which were, "Does the pressure increase or decrease with surface area?" and "What is the pressure on oblique planes?" Even after engineers had determined answers to such questions, they had to remember to account for conditions that had prompted them. In 1978, almost a century after Eiffel's concerns were current, it took a student's inquiry about the unusual support structure of the fifty-nine-story Citicorp Center to jog the mind of William LeMessurier, its chief engineer. Could oblique winds blow it over? They might have, if remedial measures had not been taken before a big storm hit Manhattan.

Eiffel had wrestled with how his Paris tower could withstand winds coming from all directions. Two years before its completion, he explained to a reporter that it would do fine, because "the curves of its four piers as produced by our calculations, rising from an enormous base and narrowing toward the top, will give a great impression of strength and beauty." Indeed, the graceful profile of the Eiffel Tower matches the mathematical curve showing that the influence of the force of the wind (its moment) to overturn the structure is greatest at its base, which is why it is broadest there.

Wind has been the bane of suspension bridges especially. In the early nineteenth century, the roadways of such bridges were regularly damaged or destroyed in storms. John Roebling studied the causes of these failures and around 1840 devised design principles that resulted in suspension bridges that could stand up to the fiercest winds. However, over the next century, bridge designers serially eroded Roebling's guidelines, believing them to be too conservative and not in keeping with the evolving aesthetics embedded in light, shallow, and narrow roadways. In the 1930s, the decks of these bridges began to move surprisingly large amounts in the wind, culminating in 1940 in the collapse of the Tacoma Narrows Bridge when the wind twisted its long, light, and flexible deck apart.

Among the notable suspension bridges built in the 1930s was the Golden Gate. It was designed according to prevailing principles of lightness and slenderness, making its roadway somewhat more flexible than in retrospect was wise. In 1951, the movement of the deck was great enough to cause some damage to the structure; within a couple of years the truss beneath the deck was stiffened by the installation of additional steel. Among the open-air, hands-on exhibits installed beside the span's San Francisco approach today is one showing the original and retrofitted trusses, complete with physical scale models that a visitor can twist to feel directly how differently they behave. Another exhibit allows a visitor to push against a plate to feel how much force it takes to resist the wind blowing at various speeds. In particular, the exhibit enables visitors to feel how the wind force increases not in proportion to speed but to the square of it. Such direct-

Scale models of a five-hundred-foot-long section of deck truss of the Golden Gate Bridge show the difference between its original open (*left*) and retrofitted closed (*right*) configuration. *From the holdings of the Golden Gate Bridge, Highway and Transportation District. Photo by Elisabeth Deir.*

contact experiences with stiffness and force obviously go beyond the words and equations an engineering student might encounter in a lecture or textbook. By manipulating these models anyone can feel not only what a force can do to a structure but also how the structure resists the force and how that resistance can be changed.

The top of the Washington Monument can move by as much as an eighth of an inch in a thirty-mile-per-hour wind. Taller, lighter, and less stiff structures naturally can move much more. In 1999, the wrought-iron Eiffel Tower swayed about four inches during a storm. When the movement is excessive for the comfort of the occupants or the good of the structure itself, measures must be taken to mitigate it. One way to do this is by means of what is known as a tuned-mass damper, in which a large mass mounted in the upper part of a tall building, say, is allowed or made to move counter to the building's motion. The principle is analogous to how we learn as children to throw our weight against the motion of a playground swing to slow down or stop it. Tuned-mass damping devices were introduced in the 1970s, most notably in New York's Citicorp Center, and they have taken many forms since. In Malaysia's Petronas Towers, once the tallest buildings in the world, heavy chains hang freely inside the buildings' spires, and when the spires move the chains are set in motion to lessen the overall effect. In Taiwan's Taipei 101, a seven-hundred-ton steel sphere painted bright yellow hangs as an architectural feature inside a five-story atrium near the top of the tower. When the building moves in the wind or during an earthquake, so does this great pendulum, which checks the overall structure's amplitude. (During a 2015 typhoon with winds in excess of one hundred miles per hour, the sphere moved more than three feet relative to the building.) Naturally, enormous forces are involved with such devices, but they should be considered backup forces that work to keep occupants of the buildings within their comfort zone.

To understand in more detail how the wind can affect a proposed bridge or skyscraper design, engineers build physical models of the structure and place them in a wind tunnel, through which air can be made to flow in a much more controlled manner than along urban streets. Among the particular behaviors engineers look for in a wind tunnel test is how stable the bridge or skyscraper model is at various wind speeds. If the model vibrates wildly or flutters at a wind speed at which a more traditional design would not, the design needs to be improved or abandoned. Sometimes, alternative designs for bridge decks or skyscraper profiles are simply compared under identical conditions. In the wake of the Tacoma Narrows Bridge disaster, which took the bridge-building community by surprise, wind tunnels were used to see if models of proposed new designs were stable enough to prevent that kind of failure from being repeated. Today, the deck design of virtually every new major suspension bridge is tested in a wind tunnel before construction begins.

When Burj Khalifa was being designed to be the tallest building in the world, there was little in its vicinity with which it could interact through the wind. However, because the structure was to be about 60 percent taller than the world's next tallest building (at the time, Taiwan's Taipei 101), engineers were acutely aware that some undiscovered phenomenon negligible in smaller structures could manifest itself in a surprising way at the new height. Thus, scale models of the building, then known as Burj Dubai, were tested to look for unusual and unexpected behavior. As the result of such tests, the helixlike arrangement of the building's setbacks evolved as an effective way to direct the wind away from forming an organized attack on the structure. By breaking up the wind into a somewhat confused pattern, the building itself was left to sit more steady amid the invisible chaos.

If the wind affects the behavior of a structure, then the structure also can affect the behavior of the wind blowing around it. In testing skyscraper models in a wind tunnel, the surrounding cityscape is often replicated in considerable detail to insure that the complexity of the interactions between wind and structures and among structures themselves is properly represented. No one wants to erect a super-tall building next to an existing tower and find upon completion that the space between the two has become a channel for the wind to blow more intensely than along some Chicago street.

17

Overarching Problems

Helping Hands

As a child I loved to play with blocks, and two objectives domi-
nated my activity. The first was stacking the blocks to make a high tower;
the second was stacking them until the tower came crashing down. Success depended as much on how well the lower blocks were aligned as on
how steady and sure my hand was in adding one more block. Regardless
of how low or high a block was in the growing tower, it was the forces
exerted by my hand that played the critical role. Setting down a single
block on a flat and level floor is a trivial task, for it matters little precisely
where the block is placed. Placing a second block atop the first is also easy;
the blocks do not have to be aligned perfectly to form a stable two-story
tower. But the experienced child knows that the better they are aligned,
the better the chances of building a high tower.

Playing with sets of perfectly cubical wooden blocks could become so
routine as to be boring. When I outgrew blocks, I looked for new experiences that challenged not only my small-motor skills and hand-eye coordination but also my mind with new and more complicated problems.
One way of achieving these objectives was to employ sets of building blocks
that were not uniformly identical in size, shape, and weight. When such a
set was not available to me, I improvised. For me, a new set of ad hoc
building blocks arrived every Saturday morning when my mother returned
from the supermarket with her weekly order. The grocery bags contained
a trove of toys disguised as bags of beans, boxes of cereal, and cans of
everything from soup to nuts. It was among the assortment of can sizes

and weights that I found my new set of building blocks and the implicit challenge to stack them to heights I had never achieved with conforming wooden blocks.

A heavy can of tomato juice was the obvious choice as the base of a tin tower. Onto it I might stack, in order of decreasing diameter, cans of peaches, asparagus, and tomato paste and top it off with a small bottle of Tabasco sauce. It was almost a no-brainer to build like this, since the bottom of one can be nested easily within the top of the one below it, and because the cans were taller than they were wide, the tower soon surpassed in height any made of conventional toy blocks. The mix of cans varied from week to week, but this alone did not present enough of a new challenge to the budding engineer who began to experiment with putting a larger can on a smaller one, thereby removing the constraint of logical order. Another variation came when the normally ignored oblong cans of sardines and tins of biscuits were also used. Instead of every tower of cans resembling a skyscraper with setbacks, in the manner of the Empire State Building, some towers suggested a building such as the one that once housed the Whitney Museum of American Art in having an upper floor jut out over a lower one. In expanding into less orthodox forms children not only hone their senses of force and balance but also may broaden their appreciation of structure and architecture.

Another challenging building activity I loved as a child was bridging a gap. Since common sets of blocks did not come in wedge shapes, making a Roman or Gothic arch was out of the question. However, flat, square, and rectangular blocks could be used to build a corbeled arch. This is one in which each successive tier juts out a bit from the one on which it rests and comes closer to meeting a corresponding one rising a distance away. The technique of corbeling employing heavy blocks of stone was known at least five thousand years ago; corbeled ceilings exist in the burial chambers of ancient passage tombs and pyramids.

The challenge to a child building a corbeled arch is first to raise its sides incrementally by stacking each block in such a way that it juts out from the one its rests upon enough, but not too much, to gain height without compromising stability. The child proceeds to do so by feeling when the weight of the block being placed causes the tower to teeter and repositioning it before adding another. Mathematicians delight in such problems as determining how many identical flat blocks can be corbeled before the structure falls over. One mathematical model says that sixty-four one-inch-long ideal blocks could extend out twenty-four inches from the base. Mathematicians may produce solutions that are perfect in theory but impractical in the shaky hands of nervous children and adults alike as they place what they think may be the last corbeled block.

The corbeled arch evolved into the more familiar true arch not by chipping away at cantilevered blocks but by using wedge-shaped blocks to achieve a graceful curve as a by-product of the construction process. Whereas straight-sided openings for windows and doors could be achieved by spanning the gap between vertical sides with a lintel, erecting an arch was not just a matter of stacking stone upon stone, since the line of stones had to reach out horizontally.

Over about four millennia, structural arches with circular, ogee, and pointed profiles were developed. Since a round or Roman arch comprising a series of voussoirs—wedges with two opposing but nonparallel sides—could not support itself until the last and topmost one—the keystone—was inserted, a timber scaffolding, commonly known as falsework or centering, had to remain in place until the arch was completed. The arch's concave lower surface naturally conformed to the convex upper surface of the falsework on which the voussoirs were assembled. After the keystone was in place, the falsework was knocked out from under the stonework, leaving the completed arch spanning the gap between abutments. In looking at a standing arch with this process in mind, we might place ourselves in the place of the workers struggling with the individual stones and feel the forces involved.

Not all arches have a prominent keystone. Many a common old stone-arch bridge has voussoirs all of roughly equal size and shape. As long ago as ancient times, construction often used brick arches above openings and hollows in walls to carry the load of upper stories around alcoves, windows, and doorways. In time, arches evolved from the semicircular Roman type, which by geometric necessity has a rise half its span, toward flatter and flatter versions. Such arches began to be used to bridge the space between iron and steel I-beams and so provide support for the floor above in fin-de-siècle buildings. The limiting case of an arch with no rise is, of course, a flat arch, in which the predominant wedge action of its voussoirs pushes strictly outward rather than downward and out.

Today, it is not uncommon to see above windows and doorways a course of bricks laid on end, with no wedge or arch action visibly apparent or implied. These false lintels are in fact typically resting on a barely perceptible steel plate. A more traditional way to bridge the span of a window has long been the post-and-lintel method, in which the horizontal lintel rests on the vertical posts that define the sides of the window opening. This look rooted in ancient Greek architecture has devolved into faux stone lintels whose length exactly matches the width of the window opening and so appear to rest on nothing but the top of the void itself, poised to slide downward like a piston into a cylinder at the slightest provocation. It is difficult to fathom, no less feel, the forces involved in such an arrangement, especially when the whole function of an arch or proper lintel in the first place was to redirect the load bearing down on a window opening to the solid walls defining it. The aesthetic evolution of architectural details in the absence of an appreciation of their historical structural roots can and does produce awkward-looking architecture. It may be difficult for non-specialists to articulate exactly what seems wrong with a faux lintel, but they may well feel that something is not quite right.

Modern stonework can also fool the eye. The Vermont stonemason Thea Alvin creates breathtaking pieces of sculpture out of flat stones laid dry one upon or beside the other. Her circular arrangements are especially striking. The individual stones are not small but are so thin and flat that they hardly evoke voussoirs. Furthermore, because the stones in a six-foot-diameter arrangement are so numerous, it is difficult to count by eye their exact number, let alone identify one as the keystone. Among the charms and fascination of Alvin's artwork is that it is a new twist on an ancient craft practiced in an ancient way. Her principal tool is a hammer, and she has given a name to every one of the dozens in her collection. Among her favorites is a seven-pounder she calls Bam Bam; a twelve-pound maul is the Convincer. Whatever its name, a hammer not only chips away at the edges of a stone to shape it to her liking but also provides the force to displace the wooden centering once the stones are firmly in place to be self-supporting.

The challenge of erecting a classical arch is the subject of a popular hands-on demonstration activity encouraged by many a so-called science museum. (Most such institutions are really science, engineering, and technology museums and laboratories, but that too-long-and-explicit description is often telescoped into what is incorrectly thought to be the all-encompassing word. The National Building Museum in Washington, D.C., is an exception. Chicago's Museum of Science and Industry found a middle ground in its very apt name.) Whatever a museum calls itself, the arch-building activity contains little science or engineering but much technology. The props for the demonstration can be small wedges of wood for an individual child to shape into an arch on a horizontal tabletop that is hinged so it can be raised up carefully to reposition the arch into a vertical plane. Alternatively, large ottoman-size wedges require building an arch to be a group activity. By raising and placing the voussoirs, usually consisting of colorful vinyl-covered blocks of relatively lightweight and slightly compressible material, the children gain a meaningful tactile experience of force. The task is simply stated: to build an arch out of the couple dozen blocks found scattered about a designated activity area. Often the blocks are numbered, indicating the order in which they are to be piled one upon another, with the starting points marked on the floor or prepared as fixed abutments. There is no wood, lumber, or timber available to construct falsework to help with the task.

Participants soon learn that achieving the goal is not as easy as it might seem. As the sides of the arch rise, they lean toward each other and the builders must keep them from toppling over. The only way the entire task can be completed is through cooperation among the construction crew. When the two arcing stacks almost meet, the keystone will not be able to be dropped into place until adjustments of the width and alignment of the gap are made. In the process of doing all this, the whole assembly is liable to collapse into a rain of bouncing blocks that settle into a jumble on the floor. Museumgoer construction teams who experience such a disaster usually want to tackle the task again, this time with the lessons learned from the failure. They now know how important the placement of each block can be, and they work toward inserting the keystone with the coordinated attention that the task requires. That done, the team can stand back to admire their achievement—until they in symbolic victory or

A family must cooperate in assembling a lightweight arch in the great hall of the National Building Museum, which is replete with masonry arches. Without cooperation, the task is virtually impossible to complete. *Courtesy National Building Museum. Photo by Kevin Allen.*

the next crew in hopeful enthusiasm knocks it down and begins all over again.

Children can learn to feel the forces inherent in arches and domes, the three-dimensional counterparts of arches, by engaging in playful games of cooperation. An educators' guide to *Building Big,* the public television series on structures and their engineering, describes a "human arch" activity in which two children, standing a couple of feet apart and facing each other, rest their palms on each other's and lean forward to form an anthropomorphic arch. In this position, the children are encouraged to inch their feet as far back as they can to better feel the forces at play. This can be dangerous if their feet slip out from under them, but if other children sit on the floor behind each of the leaning ones, they can serve as abutments or buttresses against any outward movement. Since a dome can be thought of as being generated by rotating an arch about its vertical axis, a "human dome" is formed by a modification of the arch model. A handful of kids standing in a circle facing inward and pressing against something like a soccer ball must cooperatively balance the forces their hands impress upon it. Once everyone is pushing properly inward upon the ball, they can in unison step back slightly as they raise their arms to lift the ball up above their heads. The result is a dome, and the builders feeling their hands pressing against the central ball experience the compressive forces at a real dome's apex.

Some of the grandest and most awe-inspiring interior spaces in the world are topped by structural tours de force. Standing beneath a grand dome is at the same time a humbling and an exhilarating experience. How wonderful that we dwarfed human beings can rise to such heights of accomplishment! Today in the United States, the word *dome* seems most easily to evoke images of enclosed sports stadiums, such as Houston's Astrodome and New Orleans's Superdome. Historically, however, domes have been associated not with recreational sites but with sacred places such as churches, cathedrals, and basilicas and with government landmarks such as state and national capitols. But regardless of the nature of the space beneath them, domes invariably draw the eye and mind upward in wonderment at their architectural expansiveness and engineering challenges.

Domes want to splay out at their base, but the action is more complicated than in an arch because the structural behavior is three- rather than two-dimensional. The considerable weight of a dome tends to flatten it out, which means that its edges will be pushed outward unless resisted by the massive structure on which it sits. If the resistance is not sufficient, there is a tendency for the dome to develop cracks, because as its diameter increases its circumference must necessarily be stretched. This means that the brick, stone, or concrete fabric of the dome is pulled apart, and when the pulling is greater than the material can stand, cracks develop to relieve the excessive stress. Such cracks have plagued many a famous dome.

We can get a feel for the forces and shape changes involved by considering a coffee filter or the kind of pan liner used in baking muffins. Each of these familiar kitchen accessories starts out as a flat sheet of paper in the form of a circle. The flat-bottomed shape is achieved by folding up and pleating an annular region of the sheet, which incidentally gives it sufficient stiffness to hold its shape. If one of these pleated cups is placed upside-down on a table and compressed slightly by an open palm pushing down gently on the flat center section, the pleats will open up and the edge spread out. (When the pressure is removed, the pleats will return to their prior configuration and the edge will contract accordingly, demonstrating that such a commonplace object may also be thought of as a structural spring.) Real domes are not flat topped, of course, nor are they pleated, but when they are compressed by the force of their own weight, their circumference grows and their edges spread out in a similar way, enlarging the diameter of the dome's base. Structures made of bricks, concrete, and timber generally cannot unpleat the way a coffee filter does, but if a dome is heavy enough, it will expand around its supporting edge with such intensity that the attendant circumferential stress will cause cracks to appear.

What is remarkable about the world's most notable domes is that the oldest are also among the largest. The second-century Pantheon that stands in Rome today is an all brick and concrete replacement for a temple whose timber roof was destroyed by fire. The Pantheon's hemispherical dome with the famous oculus at its apex rests on a circular cylindrical structure known as a drum that forms the building's main vertical enclosure. Although there are many penetrations of and alcoves in the drum's wall, its twenty-foot thickness provides more-than-ample strength to support the dome and resist its outward thrust.

At about 142 feet across, the domed interior space of the Pantheon has a clear span that is believed to be significantly greater than any previously built structure. Exactly how the dome was constructed is not known with certainty, but it is generally thought that some kind of falsework was employed to hold in position the forms into which the wet concrete could be

Notable domes

Dome	City	Date	Span (feet)	Material	Notes
Pantheon	Rome	AD 128	142	concrete	contains oculus
Hagia Sophia	Istanbul	563	105	brick	pendentive
Il Duomo	Florence	1436	140	brick	double octagonal
Saint Peter's	Rome	1626	137	stone	double
Saint Paul's	London	1710	112	brick, timber	double plus cone
U.S. Capitol	Washington	1866	96	cast iron	double

placed. The five-foot thickness of the dome is only slightly reduced at the coffers, which may have been cast directly into the concrete. Although these recesses may have been done to save material and weight, it is certainly likely that they were also intended for the decorative purpose that they so nicely achieve. The twenty-seven-foot-diameter oculus lightens but does not weaken the dome; the perimeter of the oculus may be thought of as a pair of semicircular horizontal arches arranged toe-to-toe to keep apart the two halves of the opening.

From the interior, the dome looks geometrically like the portion of a sphere, which is what in fact it is. From the exterior, the dome is distinguished at its base by a series of what appear to be concentric steps. These concrete features, like so much else of the structure, have been the subject of much speculation, including whether they were part of the original design. The most likely explanation for them is that they were added after the dome began to show cracking around its lower periphery, where the circumferential tensile stresses are the greatest. Adding the weight of the encircling concrete would have compressed the dome in this region, thereby closing the cracks and preventing new ones from forming. That the structure has lasted almost two millennia attests to the wisdom of the fix.

Although structurally independent of the dome and the drum, the columned front porch of the Pantheon has also been the subject of scrutiny. The original design supposedly called for a pediment supported by taller columns than the sixteen thirty-nine-foot-tall, five-foot-diameter ones that were actually used. Erecting taller monolithic columns was evidently not possible in the confined spaces around the entrance to the building, and so the shorter ones (weighing sixty tons each) were used, thereby lowering the peak of the pediment below the top of the drum. Because of this, the porch looks more squat than it should in proportion to the building proper. Technical considerations affected classical architecture, and they have continued to do so through the ages. Examples range from the introduction of flying buttresses to stabilize the walls of Gothic cathedrals to the very profile of the Eiffel Tower.

Like Rome's Pantheon, Istanbul's Hagia Sophia was a replacement for an earlier structure. The imperial capital was moved from Rome to Constantinople in AD 330, and in 360 the Church of the Holy Wisdom (Sancta Sophia) was dedicated as a cathedral. It had the form of a classical Latin basilica and was topped by a timber roof. In the year 404 a fire destroyed the structure, which was rebuilt within a decade, again with a timber roof. After another catastrophic fire in 532, the basilica was rebuilt as a more fireproof structure, and one that was considerably moreso than its predecessors. The new building was completed in about five years, and its great central dome made it a grand architectural achievement. Unfortunately, earthquakes that struck the region in 553 and 557 caused the original dome to crack, and another earthquake in 558 collapsed the dome onto the altar below. The church was again rebuilt, with a higher dome made of lighter bricks, and it was rededicated as a basilica in 562. It is essentially that structure that is known as Hagia Sophia today.

Hagia Sophia is crowned by what is known as a pendentive dome, which means that it is supported not around its entire circumference but by triangular vaulting segments (known as pendentives) that provide a geometric transition between the building's square floor plan and the dome's circular base. The basilica was further distinguished and its interior space further opened up by the semidomes that buttress its east and west ends. The physical structure continued to suffer fire and earthquake damage, as well as general deterioration, and the sacred place was often a battlefield among differing religious factions over the use of images and other theological issues. In 1453, Sultan Mehmed attacked Constantinople, wishing to convert the Christian city to an Islamic one. His siege succeeded, and the church was converted into a mosque, complete with a minaret. It was in 1935, after the Republic of Turkey had been established, that the mosque was converted to a museum, but in 2020 it was converted back into a mosque.

The Middle Ages were bracketed, roughly, by the construction of the domed Hagia Sophia in Constantinople and that of the great dome for the basilica in Florence, Italy. This latter edifice was such a great achievement that the Italian word for dome, *duomo*, came in that language also to mean *cathedral*. Indeed, Florence's Basilica di Santa Maria del Fiore is often referred to simply as Il Duomo. As was the case with many a grand cathedral or basilica, a smaller church earlier had stood on the site, and in Florence the much reworked early sixth-century structure had deteriorated and the city's population had outgrown it. A bigger and better replacement was begun in the last decade of the thirteenth century, and construction continued at a varying pace for over a hundred years, during which time several architects and master builders were involved in its design and erection. Except for its dome, the basilica was finally completed in 1418, after which a design competition was held to choose an architect-cum-engineer for the final stage. The winner was Filippo Brunelleschi, and it fell to him to design not only aesthetically but also structurally the largest dome to be built since the construction of the then-thirteen-centuries-old Pantheon.

Whereas the Pantheon's dome is hardly visible from street level, Brunelleschi's dome dominates the skyline of Florence from virtually every perspective. With no clear structural precedent—though the span and obvious constructability of the Roman dome did inspire and embolden Brunelleschi—the Florentine project was a momentous challenge and a great achievement. The superstructure is made not of concrete but of bricks, and it is a double dome, meaning that the curved and decorated surface viewed from inside the church is part of a practically separate shell structure than the one seen from the outside. The two domes are interconnected across the annular space between them and through which stairs and passageways wind to give access for maintenance and to the enormous cupola that sits atop the dome. Among the advantages of the double dome are that the overall structure is lighter and that the inner dome, whose concave surface is elaborately decorated, is somewhat insulated from the elements.

Brunelleschi's outer dome is based on a skeleton structure of vertical arches and horizontal ribs, and the spaces between them are filled in with courses of bricks (altogether more than four million). The steeply rising profile aided in building the dome without supporting scaffolding or false-work, each completed brick course being compressed into a self-supporting arch ring. According to Rowland Mainstone's seemingly exhaustive treatise on structural form, that of the Florentine dome was "much too complex to be described in detail," but its aesthetic form is easy to appreciate. The dome towers over everything else in the city, its white marble ribs standing out in stark contrast to the red terracotta infill tiles that cover so many roofs (and domes) in Florence. Atop the basilica's outer dome is the graceful and well-proportioned cupola. For it, the dome's ribs act somewhat like arch bridges, spanning the diameter of the dome and directing the weight of it all down to the dome's base, which consists of a great drum sitting upon the massive piers that border the space beneath. Indeed, it was the preexistence of the octagonal drum, which had already been erected atop the basilica proper, that constrained so much of Brunelleschi's design.

The dome crowning Saint Peter's in Rome is a structural descendant of that atop Il Duomo. However, whereas Brunelleschi had to work within the octagonal plan that had already been established, Michelangelo was able to design a truly circular structure. Because of this, his dome is structurally simpler, but the desire to have it tower over the rest of the church introduced complications that would endanger it. Wrought-iron chains were incorporated into the base of the dome in order to contain the tendency for it to splay out and thereby crack. The cracking was not fully contained, however, and it progressed even into the drum, the top of which had in turn been pushed out by the spreading base of the dome. In the eighteenth century, significant remedial work proved to be necessary.

In a twentieth-century application of dome containment and crack suppression, in excess of six hundred miles of nearly quarter-inch-diameter steel wire was wound tightly around the edge of the four-hundred-foot-diameter saucerlike concrete dome of the Assembly Hall at the University of Illinois at Urbana-Champaign—the institution's spacious indoor arena that since 1963 has been the venue for sports, entertainment, and convocation events. The wrapping process used on the Assembly Hall is a form of posttensioning, in which the tendency of the stretched wire to contract counteracts that of the dome's perimeter to expand. Without the steel wrapping, which was grouted over to hide and protect it, the Assembly

Hall's flat dome would have sought an even flatter profile, with cracks opening up in its fabric to allow it.

But cracks are not the only enemies of large structures. As is the case with the history of many churches, that of Saint Paul's in London has been one of fires. The church building traces its origins back to the early seventh century, but whether it was always in the same location remains a matter of some debate. What was believed to be the fourth Cathedral of Saint Paul was begun after the fire of 1087, and construction continued, with setbacks due again to fire, over the next two centuries. It was a very long medieval Gothic structure with a wooden roof—hence the vulnerability to fires—and a spire standing almost five hundred feet tall. This Old Saint Paul's, as it later came to be known, lost its spire to lightning in 1561, and what was left of the church was severely damaged in London's Great Fire of 1666. Even before the fire, Christopher Wren had been engaged in designing a new church for the location; the design finally settled on was in the classical tradition, which meant that its walls had to be thick enough not to require any flying buttresses. The design included a monumental dome.

Wren's dome is essentially a double one, but with an added twist. Between the inner and outer shells he incorporated a conical structure that provides support for the surmounting stonework lantern, thereby removing its significant burden from the principal shells of the double dome, which might perhaps more properly be described as a triple one. The innermost dome rises only to about half the height of the outermost one, thus enabling each to have proportions architecturally consistent, respectively, with the interior and exterior proportions of the church. The cone, which is eighteen inches thick and made of brick, has the structural advantage that the heavy load that the lantern places on it is directed by the nature of a cone's geometry straight down the inclined sides with diminishing intensity, thereby minimizing any tendency for the cone to splay out at the base and develop cracks in its brickwork. Furthermore, with the cone in place, the inner dome could be made thinner and lighter in weight than it otherwise might have had to be, and so also lower the chance of cracks developing in it.

To ensure that gaps would not open up in the brickwork, Wren incorporated a wrought-iron chain into his structural design, girdling the base of the cone with it. (The corroded iron chain was replaced with a stainless steel one in 1925.) Brick is also the structural material of the inner dome, which like the cone is only about eighteen inches thick throughout most of its shell. The outer shell of Wren's dome, which is stabilized by a network of bracing that connects it to the surface of the cone, is made of timber. It is sheathed in lead, which lends it its distinctive gray color and presumably also gives it some exterior protection from catching fire.

A fear of fire and its consequences also played a role in the design of another famous dome, that atop the United States Capitol. Architecturally, the building that is central to the legislative branch of the U.S. government evolved as the young country grew. The structure originated in a pair of small eighteenth-century buildings that by 1827 were connected by means of a construction project that included the rotunda, the congressional library rooms, and the central porticoes. By 1850, Congress had outgrown even the united and enlarged building, however, and the Philadelphia architect Thomas U. Walter was commissioned to enlarge it further. While that project was under way, it became obvious to Walter that the scale of the new wings would leave out of proportion the low brick, wood, and copper dome that Charles Bulfinch had designed over the rotunda. So, Walter designed a taller structure to replace Bulfinch's lower one.

The cast-iron dome was painted to match the rest of the building, the older parts of which were made of light-colored sandstone and the newer ones of white marble. The English critic John Ruskin believed that the use of such a modern material disqualified a structure from being considered a serious work of architecture, but his point of view has not prevailed. Cast iron was an increasingly popular structural material of the time, and its use in the Capitol dome not only made it lighter than a similarly proportioned masonry structure but also made it fireproof, something that was desirable because the old dome came close to being destroyed in 1851, when the congressional library, which then was still located in the Capitol, burned. A century or so earlier, a cast-iron structure on the scale of Walter's ninety-five-foot-diameter dome would not have been considered possible. Recall that the early eighteenth-century dome for Saint Paul's employed brick and timber as principal structural materials; not until 1779 was the one-hundred-foot-span cast-iron arch bridge erected over the Severn River at Coalbrookdale, England. The erection of the aptly named Iron Bridge not only demonstrated that each necessarily large and heavy principal iron part could be cast in a single piece but also that it could be erected into a finished structure. Still, Walter's Capitol dome weighs almost nine million pounds, and the walls of the old building had to be strengthened to bear its weight. The dome is double shelled, with a winding stairway located between the shells.

Like brick and stone, cast iron is a brittle material that is strong in compression but weak in tension. Thus, it is perfectly suited to bearing the weight of an arch or a dome comprising multiple arch ribs, but it is susceptible to cracking in regions where it is being pulled apart. Also as with a dome of brick or stone, if not properly anticipated in the design of the structure, cracking could happen between the ribs at the base and result in unsightly if not structurally dangerous fissures. The desire to avoid the cracking of domes and other structures has been a driving force in the evolution of their structural and architectural form, once again showing how important failure and the desire to obviate it have been in the development of engineering systems generally. Ultimately, the survival of our greatest monuments depends upon our understanding the forces that hold them together and those that can tear them apart.

18

Pyramids, Obelisks, and Asparagus

Ramping Up the Force

When I was a young boy growing up in Brooklyn, there was a good-size shed in our backyard that served more as a storage space than the workshop it was evidently designed to be. Unused furniture was piled in one corner, old newspapers in another, and nail kegs full of someone's bottle cap collection occupied a third. The space reminded me of the loading dock at the trucking company where my father worked as a rate clerk, and I enjoyed moving the various items from one place to another as if I were sorting them to load onto trucks that would take them to their final destination. To move a keg, I tipped it slightly and rotated the top as if I were turning a steering wheel, thus causing the container to roll on its bottom rim. The forces involved felt odd and unfamiliar. If I wasn't careful, a keg could tip over or spin out of control and bottle caps spill all over the floor. It was a lesson in heeding the feel of forces to prevent a disaster.

The newspapers were taken to the junkyard periodically to be sold at about fifteen cents per hundredweight, but first they had to be tied into bundles. For this purpose, twine and cord of all sorts were carefully un-knotted and removed from every package delivered to our home. This discarded material accumulated in the shed until there was a large enough pile of papers to secure into a bundle that was not too heavy for an eight-year-old boy and his six-year-old brother to lift. As the number of bundles accumulated, it was great fun to stack and restack them into piles that sub-divided the shed's interior with partitions that were high, but not so high that they would fall over.

On occasion, my brother and I backed our wagon up to the shed's door and loaded some bundles onto the flatbed and pulled it around the backyard. The object was to make a circuit without the bundles falling out or the wagon tipping over. When we felt especially playful we ran with the wagon around tight corners. If we were successful in returning to our starting point, we added another bundle of papers and repeated the circuit. This went on until we ran out of intact bundles. To me, it was a portent of the days I would struggle pedaling up a hill to the Heights in Queens, where my family moved when I was twelve. Then, my bike's delivery basket would be piled too high with bundles of the *Long Island Press,* and I would be fighting the forces waiting to waylay me, my bike, and our burden. It was not exactly a Sisyphean task, but it was one I had to repeat seven days a week, and with extra heavy papers full of ads and supplements on Thursdays and Sundays. Without realizing it at the time, I was engaged in the equivalent of the millennia-old challenge of moving a large piece of stone from quarry to building site and up a ramp to its destination.

The oldest known pyramid was constructed at Saqqara, Egypt, around 2800 BC. Its stepped profile distinguishes it from the more familiar plane-sided ones, such as the Great Pyramid at Giza, which was built two centuries later. Although the construction details of any pyramid continue to be debated, there is little doubt that a large number of workers had to be engaged in the project. It was thus in the engineer's interest to keep the workforce healthy and to know how to treat job-related injuries. At Saqqara that task fell to Imhotep, who is considered not only the first engineer identified by name but also the first physician. Although no direct writings by Imhotep are known to have survived, historians believe that a manuscript dating from about 1600 BC records his medical wisdom. The fifteen-foot-long scroll was unearthed by grave robbers in 1862. A 1930 translation of the papyrus showed it to contain an analysis of four dozen cases of traumatic injury of the kind expected on a construction site. What makes the cases remarkable is how modern sounding are their descriptions, examinations, diagnoses, prognoses, and treatments. Rather than following the magical and superstitious approaches to medicine that the Egyptians are believed to have practiced at the time, Imhotep took a rational approach two millennia before Hippocrates, who has long been thought to have been the first to attribute diseases to natural rather than mystical causes.

Engineers have always had to wear several hats in their work. They have had at least to learn the specialized art and science relating to whatever project they were working on at the time, whether it was the quarrying, dressing, and placement of the stone for a pyramid or the care and feeding of the workforce. In order to achieve what he did, an engineer such as Imhotep had to understand at least on a practical level how force could be harnessed to shape and move heavy blocks of stone—and how it could cause bones to break. But how Imhotep and other ancient engineers did harness the power of force is among the many intriguing open questions about ancient technology and culture: Exactly how were very heavy stones moved and the pyramids built? Many answers have been put forth, most of which have raised more questions.

The last Egyptian hieroglyph is said to have been inscribed late in the fourth century of the modern era, but serious study of Egyptian culture by Westerners did not begin until the end of seventeenth century. It was then that the first relatively precise measurements of the Great Pyramid at Giza were made. In the late eighteenth century, the French Republic sent to Egypt—under the command of Napoleon—a large expedition that included a commission to deal with scientific and artistic issues. The expedition resulted in volumes of scholarship and led to the accidental uncovering of the Rosetta Stone, which promised to be a key to deciphering the hitherto undecipherable. Thus were the foundations of Egyptology laid, and over a couple of centuries scholars and amateurs alike have built the discipline into an edifice.

As we have seen, friction always opposes motion, and it was this force that workers hauling a large building stone along a wooden or stone pathway had to overcome. According to James Frederick Edwards, a chartered consultant engineer from Manchester, England, the Egyptians probably did not construct special ramps or use incremental levering techniques to raise blocks of stone to their final resting places. Rather, he proposed a "more logical and practical alternative methodology," in which the sides of the incomplete pyramid itself were used as inclined planes up which the individual blocks were hauled on sledges. He cites experiments at the Karnak

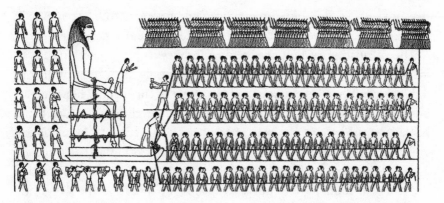

A wall drawing depicting a team of 172 men hauling a sled bearing a colossal statue of Djehutihotep enabled a modern engineer to confirm his estimate of how much force an individual worker in Egypt could pull with four millennia ago. *From Encyclopaedia Biblica via Wikimedia Commons, distributed under a CC BY-SA 3.0 license by user Newman Luke.*

temple complex that showed that "three men could pull a sledge-mounted block weighing one tonne [1,000 kilograms] over a stone surface that had been lubricated with water to reduce the effects of friction." To estimate the coefficient of friction, Edwards made an assumption about how much force an adult male can exert on a hauling rope. He took this to be 68 kilograms (150 pounds), or about 90 percent of an average man's body weight. A simple calculation then gives the coefficient of friction to be about 0.2.

To check the reasonableness of his result, Edwards looked to an ancient wall painting in the Twelfth Dynasty tomb at Deir el-Bersha. It depicts a large team of men hauling on a sledge a massive statue of the Egyptian nobleman and tomb-occupant, Djehutihotep. Using the coefficient of friction derived from the experiment at Karnak and the known weight (58 tonnes) of the Djehutihotep statue, Edwards concluded that it would have taken some 174 men to move the load. Since the wall painting shows a team of 172 men pulling the loaded sled—and we can well imagine that collectively they felt the total force with which it resisted—Edwards concluded that his assumptions and results were reasonable. He then proceeded to calculate how many men it would take to haul a building block on a sledge up the side of the pyramid at Giza, taking into account the fact that not only friction but also gravity would have had to be overcome in moving the load up one of the inclined sides of the rising pyramid.

The calculation of how much force it takes to pull something up an inclined plane is an elementary one, involving only the weight of the object, the coefficient of friction, and the angle of the incline. For the purposes of his calculation, Edwards assumed the weight of the block being hauled up the side of the pyramid under construction to be 2 tonnes, which he took to be representative of the core blocks in the structure, and assumed the same coefficient of friction as at Karnak. He recognized that the workers would have to pull also against the weight of the sledge and of the not inconsiderable length of an eight-centimeter (about three-inch) diameter rope, which he estimated at 0.3 tonnes and 0.5 tonnes, respectively. (Edwards documented such numerical details in footnotes.) Given these assumptions, Edwards deduced that it would take a team of fifty men to drag a single block up the side of the pyramid, perhaps supplemented by a few pushers to start the load moving.

He assumed that the pyramid was constructed tier by tier, which means that it rose in increments equal to the height of one core block, which he took to be about a meter and a half (five feet). Each completed level would have provided a fresh flat surface—a working plateau—to which blocks could be hauled and pushed into position to advance the structure another level. As the pyramid rose, the outer casing blocks would have been put into position on the faces up which the hauling occurred. According to Edwards, these casing blocks "would have been dressed by the stonemasons on their angled outside surfaces in order to provide a reasonably smooth surface for the blocks to be hauled up on." Oversized outer blocks would have been used, so that once all the stones were in place, the scars of construction could be erased by a final dressing.

Edwards imagined a specialized hauling team remaining on the flat top of an incomplete pyramid throughout the workday ("where they may indeed have lived during the more intensive periods of construction") to maintain efficient progress. He further posited that a number of teams would have been working there simultaneously, each being assigned to a "slipway" 5 meters wide so as not to interfere with the teams working in adjacent slipways. When the pyramid reached a quarter of its height, the 37-meter-high plateau would have been about 175 meters on a side, thus allowing for thirty-five slipways. Given the length of the slipways compared to the length of the pyramid's incline (47 meters), two teams could have worked without interference in each, simultaneously hauling blocks up two opposite sides of the pyramid and placing them on the plateau from the center out. As the pyramid rose, the working space would diminish, of course, and so would the number of teams that could simultaneously work atop it. Nevertheless, at the half height of the pyramid, Edwards estimated that by his scheme it would have taken less than three minutes to haul a block up from the base.

Of course, getting the blocks up the incline was only one aspect of the construction process. Once a block had reached a plateau, it had to be moved over to its proper place, unloaded from the sledge, the empty sledge lowered back to the ground, the rope undone and attached to a waiting sledge with another block on it, and the cycle repeated. Edwards estimated that on average it would have taken one hour to execute this cycle for each block. With multiple teams working atop a plateau, the volume of the pyramid, at least in its lower stages, could have grown faster than a block a minute. Even allowing for the complications of heavier blocks and the more complicated geometry associated with burial chambers and passages, Edwards believed that the entire pyramid (with its estimated 2.3 million individual blocks of stone) should have been possible to complete in twenty-three years, and with a workforce—including those necessary to quarry the core stone and move it to the construction site, but not including those transporting outer casing and other special stone from greater distances—at no time exceeding about ten thousand people.

But Edwards admitted that, as the pyramid approached its apex his scheme would have been increasingly difficult to implement. Reduced plateau area would have required hauling teams to work in slipways that were shorter than the incline up which the blocks had to be pulled. "Technically," Edwards conceded, "the final 10 percent (by volume) of the pyramid would have been the most difficult to construct."

Though Edwards's speculation about the great construction project does away with the need for ancillary ramps and levers to raise the blocks to their final resting height, it does not say much about the endurance or commitment of the workers. According to his scenario, each team would be engaged in a very repetitive process. Presumably, there was some relief, since an entire team of fifty would have had to be engaged simultaneously only in the actual hauling process. There could have been periods of rest for some of them while smaller teams moved and positioned the block on the plateau and others lowered the empty sledge to begin the cycle anew. But mostly it was just hard and hard-felt work.

Not all Egyptian monuments were built up stone by stone. Hence the towering statue of Djehutihotep was moved in one piece. The gargantuan task involved the application of enormously large forces by multitudes of workers pulling in unison. Obelisks presented an ever greater challenge. The classic Egyptian one was a monolith—a single large piece of stone, typically limestone or granite—with the largest intact ones approaching eighty-five feet in height and weighing several hundred tons. An obelisk is square in plan, but otherwise its geometric proportions are not set in stone. According to one survey of objects referred to as obelisks, the ratio of base width to height can vary from a relatively stubby 1:6 to a quite slender 1:12.5, with about half of the examples falling between 1:9 and 1:11. The mean of 1:9.4 is close to the commonly stated width-to-height specification of 1:10. Regardless of its slenderness ratio, an obelisk tapers gently as it rises, and its top is usually crowned by a pyramidion, whose sides are commonly inclined at a lesser angle to its base. (A pyramid itself may be thought of as a very large pyramidion or as an extremely squat obelisk.)

Carving a pillar out of solid granite was an arduous task, one that is believed to have been accomplished by repeatedly hammering the host stone with harder handheld stones, which weighed perhaps ten pounds each. The prolonged action must surely have imparted to those wielding the pounding stone a feel for the forces involved: the lift against gravity and the impact of stone upon stone. Once an obelisk was separated from the quarry, moving it to the installation site without damage and then upending it there tested the ingenuity of ancient engineers. How they achieved such a feat is not known with any certainty, but it is definitely the case that not all attempts were successful, especially when taller pieces of stone were involved. Indeed, as Galileo observed in his *Dialogues Concerning Two New Sciences,* "a small obelisk or column or other solid figure can certainly be laid down or set up without danger of breaking, while the very large ones will go to pieces under the slightest provocation, and that purely on account of their own weight." He may not have been speaking from personal experience, but he was reporting what engineers from ancient Egypt through modern times have worried about.

Some obelisks broke even before they were completely formed. Among the heaviest monoliths ever attempted to be carved out of solid rock is believed to be one commonly referred to as the "unfinished obelisk." It dates from the fifteenth century BC and remains incomplete, lying where stone carvers abandoned it in a quarry near Aswan. The size of the massive shaft is variously reported, but it would almost certainly have ended up as an obelisk approaching one hundred feet in height and weighing on the order of 750 tons. The reason the potential monument was abandoned is evidently the presence of large cracks in it. Whether these were caused by overly aggressive efforts to shape the stone—perhaps making preexisting fissures or imperfections in the material grow into full-fledged fractures—or by some other means, the flawed object was no longer suitable for its intended purpose.

Those large obelisks that did emerge from their quarry intact had to be carefully lifted and moved to their intended standing place, as Galileo had noted. But many obelisks that originated in Egypt and stood there for more than a millennium were later relocated to museums and public spaces elsewhere. The thirteenth-century BC 83-foot-tall, 360-ton monument now known as the Vatican Obelisk was moved from Heliopolis to Rome fourteen centuries later and erected there under orders of the emperor Caligula. For another fifteen centuries it stood where the Romans had installed it, but as part of the grand plan of Pope Sixtus V to construct Saint Peter's Basilica in the sixteenth century the obelisk needed to be moved about 275 yards and be re-erected in the center of the square in front of the cathedral. A competition was held to choose a plan to effect the relocation of the red granite shaft, and the scheme of the Renaissance engineer Domenico Fontana was chosen.

Fontana documented the project, which took a good year to complete, in his wonderfully illustrated book, *Della trasportatione dell'obelisco vaticano,* which was published in 1590, five years after the move. First, the obelisk had to be lifted off its old resting place, which required the construction of a wooden derrick-like scaffold and the mechanical advantage of four fifty-foot-long levers, lots of strong rope and pulley blocks—and,

It took considerable human, animal, and mechanical power—not to mention strict coordination—to relocate the Vatican Obelisk in 1586. *From Fontana,* Della trasportatione dell'obelisco vaticano.

of course, a sufficient number of workers who could apply the necessary force. When the obelisk—encased in timber for its protection—was lifted off its old base, it was rotated into a horizontal position for transportation on rollers along a specially constructed elevated ramp. The length of the journey was less than ten times the height of the obelisk. At the new location, it took seventy-four horses and nine hundred men turning forty-eight capstans in unison to raise the monolith onto its new base.

When it was time to lift the obelisk into its new upright position, the engineer demanded—under penalty of death—absolute silence from the crowd of onlookers so that the working crew could hear the leader's commands. By noon of that day, the obelisk had been inclined to an angle of about forty-five degrees off the horizontal, but the ropes began to lengthen and slip on the capstans. Suddenly, a seasoned sailor in the crowd with plenty

Significant obelisk relocations

Obelisk	Moved from	Moved to	Date	Height (feet)	Weight (tons)	Engineer
Vatican	Saint Peter's Basilica site	Saint Peter's Square	1586	83	360	D. Fontana
Paris	Luxor, Egypt	Place de la Concorde	1833	75	240	J. B. A. Le Bas
London	Alexandria, Egypt	Thames Embankment	1878	68.5	210	J. Dixon
New York	Alexandria, Egypt	Central Park	1881	69.5	225	H. H. Gorringe

of experience and feel for rigging ropes and the forces they can bring to bear, shouted, "Acqua alle funi," which translates from the Italian as "water to the ropes." The old salt knew that wetting the ropes would cause them to contract, as they did at sea, and so become functional again. When the ropes were doused with water they did indeed tighten and the procedure could safely resume. The breach of orders for silence was excused because the sailor's outburst saved the day. The monument standing today in Saint Peter's Square is the largest unbroken one outside of Egypt. In fact, Rome has the most obelisks of any city in the world, including eight from ancient Egypt and five from ancient Rome.

Other relocated obelisks traveled much farther from Egypt, and the journeys by and large took place in the nineteenth century. Among those transported the farthest is the pair—Egyptian obelisks were usually found in closely matched but not identical pairs—known by their common name, Cleopatra's Needle. One stands on the Embankment in London, and the other in New York's Central Park, not far from the Metropolitan Museum of Art. The well-known obelisk standing in the Place de la Concorde in Paris is also nicknamed Cleopatra's Needle (L'Aiguille de Cléopâtre). In spite of their attributive name, each of these obelisks predates by at least a thousand years the Greek queen Cleopatra VII, the last pharaoh of Egypt. The dramatic stories of moving these obelisks are told concisely in the splendidly illustrated book *Moving the Obelisks,* by the electrical engineer and bibliophile Bern Dibner.

The Paris Cleopatra, which stands approximately seventy-five feet tall and weighs about 240 tons, is the tallest and heaviest of the three needles, all of which demanded of the engineers entrusted with their relocation a feel for forces well out of the ordinary. The Paris needle was presented as a gift to France in 1826 and was erected there in 1833, but without its pyramidion intact. That was believed to have been damaged or stolen in Egypt twenty-five centuries earlier. In 1998, a new gold-leafed cap was finally added, serving as a reminder that the tops of ancient obelisks were often sheathed in gold or some other bright metal, which would reflect the sun as it moved across the sky. This in effect made the top of the obelisk appear to be a source of light, and so the pyramidion was believed to represent the throne of the sun god.

The sides of the Paris obelisk proper are inscribed with hieroglyphics relating to the reign of Ramses II, in whose honor it was first erected; the base contains diagrams depicting how the monolith was lowered, transported, and raised into its present position. These diagrams show that ropes, pulleys, and struts were employed to maneuver the huge stone into a horizontal position, with oak logs fitted to one edge of the base constituting a hinge that was also supposed to protect that part of the stone from damage. A barge, named the *Luxor* (after the obelisk's original home in ancient Thebes), was constructed and beached as close to the stone as the receding Nile would allow. Once the monolith was aboard, the opening in the barge's bow was closed, and when the river rose, the vessel was allowed to float toward the river's mouth. There, the steamship *Sphinx* began towing the barge and its massive cargo across the Mediterranean, along the European coast, around the Iberian Peninsula, into the English Channel, and up the Seine. The barge had been designed with the proper draft and height to clear both shallow waters and low bridges. From the floating of the barge in Egypt to its arrival quayside in Paris took two years. The obelisk was hauled off the barge and up a ramp to the Place de la Concorde by five sets of compound blocks-and-tackle and capstans, each operated by forty-eight men. (The use of a steam engine had been planned, but the still relatively new machine did not prove reliable.) The stone was righted in France in a process reverse to that by which it had been lowered from its long-standing perch in Egypt.

The other two Cleopatra's Needles, which date from the fourteenth century BC, were a pair that stood in Heliopolis until, during the reign of Augustus Caesar, they were moved to Alexandria, where they served as Roman monuments. One of two early fourteenth-century AD earthquakes is believed to have toppled one of the obelisks, and as it lay on the ground it was gradually covered over by sand. After the 1801 Battle of Alexandria, British army officers decided to take the supine obelisk as a commemorative trophy. All that had to be done was to figure out how to transport it to England. The original plan was to use a sunken French frigate, which was raised for the purpose, but it was washed away in a storm. Decades passed with no real progress being made toward moving the obelisk, which was almost seventy feet tall and weighed about 210 tons. It was not until 1877—when the civil engineer John Dixon, who had widespread experience in the construction of railways, bridges, and ports, was commissioned to move the monolith—that things really began to happen. The agreement with Dixon called for him to get the great stone to London and erect it on a bank of the Thames—for a fixed sum that did not allow for all of the ways in which his scheme could go wrong. This oversight, combined with some bad luck, would take its toll. The final cost of moving the obelisk has been estimated at four times the $50,000 he was paid to accomplish the task.

Dixon's scheme called for using modern machinery, including, in addition to steam engines, hydraulic jacks and seaworthy iron vessels. However, instead of the obelisk being dragged into the hull of a fully formed ship, a hull was assembled around the stone while it remained in its sandy resting place. The hull was prefabricated in sections at a London shipyard and sent to Alexandria for assembly. The ninety-three-foot-long, fifteen-foot-diameter cylindrical structure was designed with ten watertight compartments and timber cushioning to hold the obelisk snug in rough seas. After it encased the monument, the pontoon could be rolled (under controlled conditions) to the waterfront, launched, and outfitted with a keel, a cabin, a bridge, masts, and sails. There remained the small detail of demolishing a seawall that stood between the land and water, but that was accomplished with the aid of dynamite.

The launching went more or less according to plan, but the best-laid plans are often thwarted. It was soon discovered that the hull had been pierced by a sharp stone and that, because bulkhead doors had inadvertently been left ajar, water flooded several compartments. The hull had to be repaired and water pumped out before ballast could be added and the *Cleopatra*, as the unusual vessel was unimaginatively called, could get on her way. In September 1877, she began to be towed by the steamer *Olga*, but it was soon clear that the *Cleopatra* was not the most seaworthy of vessels. She pitched and yawed excessively, and a storm redistributed her ballast, causing her to float on her beam. Sailors were dispatched to stabilize the ballast, but they and the *Cleopatra* were believed lost. The *Olga* sailed on to England without its cargo. Fortunately, the watertight pontoon actually survived the storm and was towed to port by a ship that found her floating free. The obelisk in its curious vessel finally reached its destination in January 1878.

The obelisk was to be erected on the Thames Embankment, where a new pedestal had been prepared for it. To release the stone from the iron coffin that had kept it from sinking, the pontoon was positioned at high tide on a timber crib. Once in place, the vessel was disassembled, leaving the obelisk bared in a horizontal attitude. It was raised by means of hydraulic jacks to a proper height where it could be pushed by screw jacks over the waiting pedestal. To rotate the monolith into a vertical position, a massive timber frame was employed, it being designed to take advantage of the kinematics and dynamics about the stone's center of gravity. The job was completed in September, almost eight months after it began. Almost three years to the day after the obelisk had left Alexandria, Cleopatra's Needle stood tall on the Thames Embankment, the first time in almost six centuries that it once again pointed up to the sky.

Initially, the London obelisk was set directly on its base, but it looked unstable to some observers. The impression was heightened by the rounding of the monolith's bottom edges and corners, a common result of transporting, erecting, and relocating any large piece of stone. In some cases, the rounding resulted when an obelisk fell off its pedestal. To make such a monument appear more stable—and in fact in some cases to ensure stability—footlike objects were wedged under the corners. Stone or bronze supports are believed to have been formed in the shape of cubes, balls, skulls, lion's feet, and crabs, these last being related to the scarab, which to ancient Egyptians symbolized immortality. In the case of the London obelisk, the condition of the base was concealed behind decorative bronze castings.

The third Cleopatra's Needle, which has become known also as the New York obelisk, is about a foot taller and fifteen tons heavier than the London one. It had stood erect in Alexandria until 1879; a little more than a year later it did so again on a pedestal in New York's Central Park. This obelisk has a bronze replica of a crab weighing nine hundred pounds wedged under each corner, but the extensive rounding of the corners remains clearly visible. In the course of being moved, it had to be lowered to a horizontal position, transported to a ship, moved through a sea and over an ocean, hauled halfway across the island of Manhattan, and, finally, raised to the vertical atop its pedestal set behind the Metropolitan Museum of Art. In the tradition of Domenico Fontana, U.S. Navy lieutenant-commander Henry H. Gorringe, the engineer of the move, documented the achievement.

The obelisk that stood in Heliopolis for almost fifteen hundred years before being transported to Alexandria and, much later, to New York, is covered with hieroglyphics on all four faces of its main shaft and also of its pyramidion. These inscriptions record that the monolith was created and first erected to honor the pharaoh Thutmose III, who reigned during the fifteenth century BC. The obelisk also bears the name of Ramses II, whose dynasty occupied much of the thirteenth century BC. A third pharaoh, Osorkon I, from the ninth and tenth centuries BC, is also memorialized on the stone. The relocation of the ancient monument from Alexandria to New York was underwritten by Commodore Cornelius Vanderbilt's eldest son, William H. Vanderbilt, who at the time of his death was believed to be the richest man in the world. But rich men do not get that way by spending their money carelessly. Vanderbilt's underwriting of the estimated $75,000 that the move would cost was to be paid to Gorringe only when the obelisk was safely relocated to Central Park.

Archaeologists opposed the removal of the obelisk, but since it had officially been given by the Egyptian khedive to the United States, they could do little but refuse to cooperate in moving it. However, Gorringe was able to accomplish the task efficiently without them. His scheme involved erecting steel towers on masonry piers that stood on opposite sides of the obelisk. The towers were topped by trunnions located at the height of the center of gravity of the monolith, which had been sheathed in timber and fitted with a steel collar around the location of balance. With this arrangement, the obelisk could be rotated from a vertical to a horizontal position. Once there, stacks of timber were erected under each end of the obelisk, and it was lowered to the ground in stages by sixty-ton hydraulic jacks working alternately beneath points near the shaft's base and its top. To keep the long piece of stone from bending excessively during this critical operation, its casing was fitted with a "steel rope truss," which helped support the cantilevered half of the obelisk when a hydraulic jack was being repositioned. The process of lowering the stone into a caisson that would float it to an awaiting steamship took only six months from the time the equipment arrived in Alexandria.

The steamer *Dessoug* was located in a dry dock about a mile from the obelisk, but Gorringe could not secure permission from Alexandrian authorities to move the trophy through the city. This meant that it had to be floated in its caisson around a jetty, which required taking about a ten-mile detour to reach the ship. Once this was done, the obelisk was deposited on a timber platform erected around the ship and dragged into its hold through an opening that had been made in the hull. Once the main cargo was in place, the hull hole was closed to make the ship seaworthy. The monolith's fifty-ton pedestal and foundation stones were loaded onto the ship through a deck hatch by means of heavy-duty cranes. Timber shoring and sheathing were installed around the cargo to keep it from shifting at sea. From the time the obelisk was floated into the dry dock to the time it was ready for the sea journey was about six weeks.

After about five weeks at sea, the *Dessoug* sailed up the Hudson River and docked at West Fifty-First Street. The pedestal and foundation stones were removed first and transported to their destination. By the common New York City distance measures of long and short blocks, this amounted to about six long and thirty short city blocks to the top of a knoll behind the Metropolitan Museum. The site had been prepared down to Manhattan's famous granite bedrock, which had been leveled to receive the foundation stones upon which the pedestal would be placed. Moving the individual stones was not remarkable, but moving the fifty-ton pedestal was a challenge. As had been necessary to load it into the ship in Alexandria, two cranes operating in unison were required to lift the stone out of the ship's hold. It was transported through city streets by a thirty-two-horse-drawn carriage, from whose frame it was slung. When the heavy load caused the carriage wheels to get stuck in ruts from which the horses could not extract it, hydraulic rams were called into service.

Moving the obelisk proper was even more difficult, and attempts by dry dock owners to realize excessive profits further complicated the task. To avoid paying outlandish rates, the *Dessoug* was sent to Staten Island, where its hull could be opened and its unique cargo extracted at more reasonable rates. At first, an old method of rolling the burden on steel cannon-balls was employed, but the concentrated weight caused damage to the channels in which the balls traveled. This led to a change in plans, and rollers riding on steel rails were employed to move the obelisk onto a timber pontoon, which floated the monolith to a pier on West Ninety-First Street, somewhat closer to the museum but still a good six long and ten short blocks away from it. Traveling the two miles between the pier and the awaiting pedestal involved crossing a major railway and dealing with elevation changes of as much as 230 feet. The journey was accomplished with the aid of a small steam engine mounted on the same platform as the obelisk. A block-and-tackle system connected the engine's drum to an anchor located some distance ahead, and the engine pulled the platform over rollers traveling on timber, which could be repositioned after the burden had passed over them. From the pier, it took 112 days to reach the final destination. Once there, the horizontal obelisk was rotated into a vertical position in just the reverse way that it was taken down in Alexandria. The final placement on the pedestal occurred in January 1881, only sixteen months after the task of moving it had begun. This pleased William Vanderbilt so thoroughly that he paid Gorringe the full relocation cost of $104,000.

The modern movement of obelisks has been well documented, but how they were raised originally has been called "one of the most intriguing engineering feats of the ancient world" and, compared to the plethora of scholarship and speculation on the construction of pyramids, has largely been overlooked. In order to bring some attention to the problem of the obelisks, the Public Broadcasting Service's science television series *Nova* undertook in 1994 a project to move and put upright a forty-three-foot-long, forty-ton obelisk using only the power of human beings pulling in unison on ropes. The slow-going attempt was abandoned in deference to the television production schedule. A second attempt made in 1999 also failed, but later in that year a twenty-five-ton obelisk was successfully raised in a Massachusetts quarry, thus demonstrating, albeit on a relatively small scale, the validity of a technique that relies on a sandpit. Of course, even such a successful demonstration does not prove that this was the way the ancients accomplished the task, but accomplish it the quarry crew certainly did, and the arduousness associated with moving obelisks even with modern equipment makes the accomplishments of the ancients all the more admirable.

Not all difficult tasks involve exerting great forces on large, heavy things. There is an episode of the television sitcom *The Big Bang Theory* in which the character Sheldon feigns struggling to open a jar of white asparagus, which gives him an excuse to ask his roommate Leonard to open it and so impress upon his date, Stephanie, that he is the alpha male. Leonard utters real grunts of effort and frustration at finding the task difficult. After trying unsuccessfully to twist the lid off, first by holding the jar out in front of him and then against his body, he gives it a gentle tap against the edge of the counter, presumably hoping that this will somehow loosen the stubborn lid. When it does not, Leonard gives it a harder knock on the counter, and the glass jar shatters and cuts his thumb. Stephanie, who is a surgical resident, takes him to the hospital and stitches up the wound.

If Leonard the experimental physicist had reflected on the forces acting between the jar and its screw-on lid, he might have demonstrated himself to be Sheldon's superior and impressed Stephanie. What did he not understand? Unfortunately, a television sitcom, like a science fiction tale, is not always true to the real laws of nature. Had the engineer character Howard been present he might have advised Leonard to apply his knowledge of physics—in the form of engineering science—more effectively to opening the jar. Howard might have wisecracked, "It's not science fiction; it's science friction. Let's look into it."

Asparagus and other vegetables can be preserved in a jar for years. The canning process involves first heating the food contained in a jar with a finger-tightened lid to the temperature of boiling water. This not only kills bacteria but as everything cools down creates a tight seal because of the pressure difference between the contents and the atmosphere. This can be verified by pressing down on the center of the lid: If it does not give, the contents are effectively vacuum sealed. Some lids have a so-called button in the center, which in a depressed position, provides a visual confirmation that the contents are "vacuum packed." However, if any harmful microorganisms survived the heat of canning or if the seal fails to keep air out, the pressure inside the can will increase and the button, which can be thought of as a kind of spring or jumping disk, will snap up. This is why consumers are advised to discard any can or jar with a raised button.

For an unspoiled jar of asparagus, the forces that keep the lid tightly on are essentially the same ones that can make it so difficult to open. Whether a particular jar will be hard or easy to open can depend on the details of the geometry where the two parts meet, but in general the bottom surface of the internal screw threads of the lid will be pulled down onto the top surface of the jar's external screw threads. When forces (in the form of a couple) are applied to the lid, friction forces along the surfaces of contact between the threads and between the annular part of the lid's underside that is in contact with the top of the jar's rim will act in opposition to the twisting forces. Since the goal is to unscrew the lid so that it can be lifted off the jar, our instinct is not to push down on it while twisting. The harder the lid is pressed down, the more friction will develop between it and the jar, and the task will grow more difficult the greater the effort. But if we do not push down with our palm atop the lid while our fingers grasp its periphery, we will not receive any increased benefit from the friction force between our palm and the lid. Since the coefficient of friction between skin and steel is greater than that between glass and steel, the counterintuitive downward push will help us in freeing the lid.

Unscrewing stubborn bottle caps and lids of all kinds is a common problem, and many approaches have been devised to solve it. Some people find the screw tops on wine bottles to be especially difficult to remove. One cause of the difficulty is inadequate perforations between the metal collar and top proper, so they rotate together about the neck of the bottle rather than separating. One wine lover solves the problem by severing the ligatures with the blade that is often incorporated into the same jackknife containing a corkscrew. Other people find it difficult to apply enough twisting force to the small diameter of the top. Catherine learned an easy workaround from one of her friends. Instead of holding the bottle still while trying to twist off the top, her friend holds the top in place and twists the bottle, the larger diameter giving her a greater mechanical advantage.

I deal with tough wine- and beer-bottle-opening chores differently. With my right hand I do hold the bottle by its neck in the usual way, meaning how I might do so to lift an open bottle to my lips for a sip. However, I do not place my left hand around the cap as if I were choking up on a baseball bat. Rather, I turn that hand around so that its thumb abuts the one on my right hand. Now, when I begin to twist the top counterclockwise, it rolls into rather than out of the crook between my thumb and index finger, and so as I turn my hand my grip tightens the way that of a monkey wrench does to unscrew a tightly coupled pipe.

Bottled carbonated beverages present an additional problem. Since the pressure in the bottle will continue to push the cap upward and maintain the contact and friction forces between mating threads, the pressure will be effectively contained until the cap is almost totally unscrewed. If the cap is being turned very quickly, the unrelieved pressure can cause the cap to fly off and possibly hit someone in the eye. To avoid this, the threads on both the bottle and cap are typically not made continuous. Rather, they are suggestive of helical ramps with gaps, not unlike the broken center line on a two-lane road that indicates where passing can be done safely. Discontinuous bottle-cap threads provide a way for pressure to pass out of the bottle as it is being opened. The hissing sound accompanying the opening is signaling this.

19

Moving with the Planet

Feeling the Earth Quake

Buildings stand up because engineers imagined how they could be knocked down. As strange as this may sound, the surest way to guarantee that any structure—building, bridge, tower, dam—does not collapse is to try to make it do so, but not necessarily in real life. It would be silly and expensive to build an actual skyscraper, say, without having a good idea that it would stand up to whatever forces might attack it.

All manmade structures can come under attack by the forces of nature, which include gravity, earthquakes, hurricanes, and tsunamis. These are not things that engineers can easily create or duplicate, but they can construct models of them—virtual models that do not even require real concrete, steel, wind, or water. Engineers know whether a building will stand up because they attack models of it in their minds, on paper, on the computer screen, or even in a laboratory before construction ever begins. The laboratory models are digital versions of the real thing, but they are not exactly computer games. Every child is familiar with models. From an early age, boys and girls playing with real and virtual building blocks and other materials construct real and imaginary arches, bridges, houses, castles, towers. We call these things toys, but they teach us a great deal about how to imagine, create, and assemble parts into a whole. And they help us develop a feel for the forces involved.

Older generations may remember making models of structures by connecting with little nuts and bolts small steel parts that came in Meccano and Erector construction sets, or with wooden parts that came in sets called

Tinkertoys and Lincoln Logs. Today, children use plastic blocks known as Legos. As anyone who has played with any of these building sets knows, there are limits to how large an object can be made. Sometimes we run out of parts; sometimes we use the parts too boldly and our construction collapses before we can finish it.

Real engineers are not playing when they design a structure, but their physical models can collapse all the same, by being too heavy, too unbalanced, or just too vulnerable to withstand the forces induced by heavy steps of a colleague that make the floor shake like the ground does in an earthquake. Sometimes, engineers build more sophisticated models on large mechanical devices called shake tables so that they can deliberately move them to imitate an earthquake. When collapses occur, children and engineers alike learn from the failure what to watch out for the next time they build. Long ago engineers learned to reinforce walls by laying bricks in crisscross fashion or tying them together with strips of iron. Modern masonry structures often incorporate steel rods passing through hollow cinderblocks to provide strength against a wall of them being too easily pushed over. Modern engineers learn from digital models when and how to use larger and stronger beams and columns.

Like children playing with toys, adult engineers learn from failures what does not work, what has not been able to stand up to the forces man and nature can throw at them. And by knowing what has not succeeded in the past they know better how to design things that will in the future. This is the way structures have been improved and perfected since the days of the ancient pyramids. Then, knowledge was accumulated largely by means of trial and error. Today, buildings stand up because modern engineers try to knock down conceptual, pencil-and-paper, and virtual models of them. The process is known as design by analysis, and it represents a careful thinking ahead about what can go wrong. But there are always surprises.

The unusual earthquake that shook the East Coast of the United States in August 2011 allowed residents there to experience some unfamiliar forces and left its mark on some remarkable structures. In Washington, D.C., National Cathedral spire sections shifted, finials toppled, and the grounds were left littered with fallen stone angels. Repairs were expected to cost millions and take years to complete. The late-nineteenth-century pension building, which now houses the National Building Museum, was closed until the enormous brick structure with its spectacular interior space could be inspected. It was declared safe, but models of the Empire State Building and Burj Khalifa, which were part of the museum's exhibit of Lego architecture, suffered the collapse of some of their topmost plastic brickwork. The actual Washington Monument developed several cracks in its stone facing, but these were just the latest insult to a structure whose realization was as rocky as it was prolonged.

The Washington Monument is often described as an obelisk, and sometimes even as a "true obelisk," even though it is not. As we have seen, a historically true obelisk is a monolith, a pylon formed out of a single piece of stone, solid all the way through. The monument in the U.S. capital is built up of individual blocks and so is properly described as a masonry structure, as is any building whose basic construction materials are bricks, cinderblocks, and the like, whether joined together with mortar or just piled dry. The Washington Monument, which contains blocks of marble, gneiss, granite, and sandstone, is believed to be the world's tallest unreinforced masonry structure, meaning that it has no iron or steel rods or clamps among the stones. At any one time, the weight of all the people inside the Washington Monument, which weighs about eighty thousand tons, probably amounts to less than ten tons. In other words, the monument barely feels their presence. However, the magnitude of the task of lifting all the stones into place, and the weight that must bear down on the foundation, might easily be imagined by observing the massively thick walls that must be passed through to gain access to the confines of its interior. The experience can be heightened by recalling the story of the Washington Monument's planning, design, and construction, which spanned almost a century and, not surprisingly, considering its significance and location, was embroiled in political, financial, and other nontechnological controversies. But an unadorned obelisk, whether true or not, was not the only idea for a monument to our country's first president.

A memorial to George Washington in the form of an equestrian statue was proposed as early as 1783, after the War of Independence had been officially ended by the Treaty of Paris. A committee of the Continental Congress reported favorably on the proposal, and Congress resolved specifically to erect a bronze statue of the general wearing Roman dress and holding a truncheon in his right hand. The statue, which was to be located "in the place where the residence of Congress shall be established," that is, the nation's capital, was to be mounted on a marble pedestal, around which in bas-relief would be represented the "principal events of the war, in which General Washington commanded in person." The congressional resolution further specified that the statue be created by a European sculptor, who would be provided a likeness of Washington and a description of relevant events of the war. The offshoring of the monument was driven not by economics, as it might be today, but by the belief that there were few if any artists in late-eighteenth-century America who had the talent or skill to execute a work of such importance to the young country.

A specific location for the new nation's capital had not yet been chosen, let alone the precise location for the statue of Washington. Indeed, there was still the matter of drafting and ratifying a constitution. That happened in 1787, of course, and the decision to locate the capital beside the Potomac River was reached in 1790. Responsibility for the plan of the city would fall to Pierre Charles L'Enfant, who was born in France but came to America to fight in the Revolutionary War. Following the war, L'Enfant went to New York City, where he established a civil engineering firm. Consistent with his earlier training at the French Royal Academy in the Louvre, L'Enfant also gained a reputation in architecture and design, so when he expressed to President Washington an interest in designing the new federal capital, his proposal was taken seriously. L'Enfant was given the responsibility of preparing drawings of street layouts at suitable sites for the city and the public buildings that would be located in it. His task of siting structures was the classic architectural one that Vitruvius had written about regarding ancient cities. L'Enfant took his job to heart and went far beyond what was expected of him.

L'Enfant's 1791 plan for Federal City, which was to become known as Washington, D.C., is well known, and the layout of its streets and landmarks today closely follows the most salient features of the original one on paper. In particular, the Capitol sits atop a hill and the President's House, as the future White House was then called, atop a ridge. These houses of the Congress and the president are about a mile apart along Pennsylvania Avenue. The National Mall extends westward from the Capitol toward the

Potomac. The equestrian statue was part of the plan, and it was to be located at the intersection of an east-west axis through the Capitol and a north-south axis through the White House. This is, of course, approximately where the Washington Monument now stands. Why it is not an equestrian statue and why it is not exactly where L'Enfant located a monument to George Washington is part of the story, every aspect of which involves force in one form or another. Naturally, there were the physical forces of pull and push necessary to overcome gravity and friction, but the far greater impediments to progress came in the form of cultural, aesthetic, political, and economic pushes and pulls.

One of the reasons that plans for the grand equestrian statue did not proceed with alacrity was that Washington himself objected to a memorial while he was still living. After reluctantly serving a second term as president and then refusing to serve a third, Washington retired to his Mount Vernon home in 1797 and died there two years later. With no progress having been made on an equestrian statue, Congress appointed a committee to make a recommendation for honoring the man who would be called the father of our country. Within ten days of Washington's death, the committee recommended the creation of a marble monument, which would "commemorate the great events of his military and political life" and in which it was hoped the first president's body would be interred. A resolution by Congress specified that the monument be located inside the Capitol, which was yet to be built. The idea languished for decades. In the meantime, construction of the Capitol began, and a mausoleum was included under the rotunda, there being an expectation that not only George but also Martha Washington's remains would be interred there. That did not happen, of course, and Mount Vernon continues to be the Washingtons' resting place.

The year 1832 marked the centennial of George Washington's birth, and this occasioned renewed efforts to memorialize the first president. Congress authorized $5,000 to be spent on a marble statue that would stand in the Capitol rotunda. The American sculptor Horatio Greenough was commissioned to create the likeness and produced a loosely draped figure that was bare-chested. The statue was ridiculed as making Washington look, among other things, as if he were entering or leaving a Roman bath. It was not installed inside the Capitol but rather relegated to the grounds outside, where it suffered the indignity that birds can inflict upon public monuments. Eventually it was moved into the Smithsonian Institution.

Citizens interested in memorializing George Washington in a respectful way began to organize efforts independent of Congress. The Washington National Monument Society was formed in 1833 and elected Chief Justice John Marshall its president. In spite of tough economic times, the society launched a campaign to encourage every American citizen to donate one dollar toward a monument that was expected to cost close to $1 million. After about three years, only $28,000 was collected from a population of about fifteen million, but this was enough to sponsor a design competition. Entries ranged from Victorian towers suggestive of the monuments to Prince Albert and Sir Walter Scott that stand today in London and Edinburgh, respectively, to what looked like an Egyptian obelisk rising out of the center of a circular colonnade. This latter was the creation of the American architect Robert Mills, who was also the designer of the 178-foot-tall white marble Doric column monument topped by a statue of George Washington clad in a Roman toga that had been erected in Baltimore while Congress dithered.

Mills considered his design for the capital monument to be "a national Pantheon," containing as it did spaces behind the columns for installing statues of prominent Americans. But critics of the six-hundred-foot-tall obelisk within the one-hundred-foot-high colonnade saw it as a hodgepodge of Babylonian, Egyptian, Greek, and Roman styles. Nevertheless, it did win the competition, although fortunately Mills's original design was not the one that was finally erected. Continuing difficulties in raising money led to the scrapping of the colonnade and to the reducing of the height of the obelisk from six hundred to five hundred feet, perhaps to be embellished upon later. The intent was to locate the monument at the intersection of axes through the Capitol and through the President's House, as L'Enfant had planned, and Congress granted thirty-seven acres of land containing that location for the purpose. However, after the soil was tested, the area was found to be too marshy to support the weight of the monument, so a site one hundred yards to the southeast was selected, a decision that upset somewhat the alignment that L'Enfant had envisioned.

Unlike a monolithic obelisk, the very tall Washington Monument could not be made in a horizontal position and then lifted into place. Building

An 1833 proposal from the architect Robert Mills imagined a Washington monument enclosed by a colonnade.
Many iconic structures end up looking different from
the way they were first conceived. *Library of Congress,
Prints & Photographs Division, LC-USZ62-51521.*

the monument piece by piece was to be a long and difficult process, something not foreseen when the cornerstone was laid on July 4, 1848, and
construction commenced. The twenty-four-foot-deep foundation consisted of large blocks of blue gneiss set in lime mortar and cement mixed
together. The base measured eighty feet square, and the orderly pile of
stones rose in the shape of a truncated stepped pyramid that at ground
level was a bit larger than the base of the obelisk.

The superstructure was built of blue gneiss stones of irregular shape set
in mortar and faced with marble on the exterior and interior walls. The

shaft was designed to taper from fifty-five feet square at its base to thirty-four feet square at the five-hundred-foot mark, and because the walls had to support a lesser and lesser burden as they rose, they necessarily grew thinner to maintain an ample interior space. This void would eventually accommodate a centrally located elevator shaft and stairway between that and the monument's load-bearing walls.

Construction continued until 1854, when the Monument Society's building funds became exhausted. In 1855, with the structure standing about 150 feet tall above its foundation, Congress appropriated $200,000 to keep the work progressing, but that money was eventually rescinded in the wake of a hostile takeover of the Monument Society by members of the Know-Nothing Party, who feared that the growing number of German and Irish Catholic immigrants would lead to control of the country by the pope in Rome. During their takeover of the project, the Know-Nothings added some height to the structure, but they used inferior masonry that had previously been rejected for use in the monument. With the decline of their party, the Know-Nothings left the project in 1858, and the unfinished stone stump would stand at about 156 feet high for two decades as a national embarrassment. It was the approaching celebration of the centennial of the Declaration of Independence that prompted Congress to do something about the situation, and a law was passed whereby the U.S. government took over the project. In 1876, the incomplete structure and the land on which it stood was ceded by the Monument Society to the government, and the U.S. Army Corps of Engineers was made responsible for completing the obelisk. Lieutenant Colonel Thomas Lincoln Casey was put in change in 1878, and construction resumed in 1880.

The first order of business was to reinforce the original foundation. The weight of the incomplete structure bearing down on it was already in excess of thirty thousand tons, and the monument had reached less than a third of its final height. In order to raise the capacity of the foundation, its depth was increased to 36 feet and its spread was increased to 126 feet square. This was done by undermining the original foundation in stages and infilling the spaces with concrete, thereby significantly strengthening the base. In the superstructure, not only was the stonework done under the Know-Nothings inferior to begin with but also its top courses had been damaged by being exposed to rain, snow, and repeated freezing and thawing over the years of construction inactivity. Casey ordered 6 feet of marble removed and reset with new stone that would restore the geometry of the monument to its proper state and provide a sound base upon which to resume work. The marble that had been used from the beginning of construction had come from a Maryland quarry, but marble from Massachusetts was used for a while after construction was taken over by the Corps of Engineers. When the color was seen not to match, the builders returned to the Maryland quarry for the facing material.

At its base, the walls of the monument had been made 15 feet thick. As they rose, they thinned; at the 500-foot level the walls are only eighteen inches thick. Up to the pylon's 452-foot level, there is white marble outside but dressed gneiss granite inside the walls; beyond that level, the walls are marble throughout. The design of the top of the monument soon came under scrutiny. Mills's original plan had shown an "obelisk" topped by a rather squat pyramid, but in the meantime George Perkins Marsh, the first U.S. ambassador to Italy, had done some research into true obelisks, which he found were supposed to be ten times as high as they were wide at the base. Since the base of the rising monument had been established at a bit more than 55 feet, this meant that the final total height should be 555 feet. In order to achieve this height, the shaft would have to be topped by a rather pointed pyramidion 55 feet tall. Colonel Casey had originally planned for the squat pyramidion, and he imagined one of metal with glass inserts to let light into the monument's interior space. However, in consultation with Bernard Richardson Green, a civil engineer member of Casey's staff who would go on to work on the Library of Congress building and its book stacks, a metal roof was rejected because it would be too heavy for the thin walls near the monument's top; it would also be prone to rust, and the runoff from the roof would likely cause the white marble walls to discolor. The top of the obelisk would therefore be made of marble, too. Structurally, it was designed to be made of marble slabs resting on marble ribs, and totaling about three hundred tons in weight.

Just as an arch is finished with a keystone, so a pyramid is with its capstone. The one atop the Washington Monument weighs 3,300 pounds, and putting it in place took careful planning and staging. Since the monument was designed to contain an observatory, its pyramidion has eight windows near its base. These also served an essential purpose during construction, for putting the finishing touches on the monument required a series of scaffolds to which access was gained through the windows. Once the capstone was set, the only thing that remained to finish the exterior of the monument was to add a metal tip to the very top to serve as a lightning rod. Aluminum, which was a scarce and expensive metal at the time, was chosen because of its high electrical conductivity. Furthermore, because pure aluminum did not tarnish in the air, water flowing off of it would not carry residue that could stain the marble. The one-hundred-ounce aluminum tip on the Washington Monument was the largest piece of that metal cast to date.

The nine-inch-tall aluminum apex is inscribed on all four sides. The north side, which faces the White House, contains the names of the joint commission created by the 1876 act of Congress to oversee the completion of the monument. The list is headed by Chester A. Arthur, U.S. president at the time the capstone was set. On the west side of the aluminum tip are three significant events and dates: the cornerstone laying, on July 4, 1848; the resumption of construction with the first stone laid at 152 feet, on August 7, 1880; and the setting of the capstone, on December 6, 1884. On the south face of the tip are the names of the chief engineer and architect, Colonel Casey, and his three assistants, including Bernard Green. The east side of the aluminum tip is inscribed simply "Laus Deo"—"Praise be to God." The monument was dedicated on February 21, 1885, the day before Washington's Birthday, which fell on a Sunday that year.

It remained to convert the interior from one resembling a construction site to one welcoming to visitors. The wooden planks that had served workers had to be replaced with iron stair treads—898 in all—and handrails added. A passenger elevator had to be installed, albeit one that would take five minutes or so to rise or descend the five hundred feet to the observation level. Employing a new technology for the time, an electric lighting system consisting of seventy-five incandescent bulbs also had to be put in place, with the dynamo powering them to be located in a separate engine house. There were also memorial stones to be set into the wall beside the stairway. These were gifts from states, societies, foreign governments, and the like that had been solicited by the Monument Society for setting into the interior walls. When it was experiencing difficulties in securing cash gifts, the society encouraged the donation of such stones, which reduced the number of building blocks that had to be purchased. A monument stone that had been a gift from the Vatican and that was referred to as the Pope's Stone was spirited away by the Know-Nothings during their reign over the construction site. The stolen stone, which was supposedly thrown into the Potomac, was never recovered, but a replacement for it was secured in 1982.

The Washington Monument opened to the public in 1888. It was not without its critics, one of whom regretted the strengthening of its foundation because now it would be ages before the obelisk would fall down. Others lamented the fact that Robert Mills's original plan had not been followed more closely, believing that without the colonnade the monument was "an ugly chimney." The monument also had its admirers, including the landscape architect and conservationist Frederick Law Olmsted Jr., who thought that it stood "not only as one of the most stupendous works of man, but also as one of the most beautiful of human creations. Indeed, it is at once so great and so simple that it seems to be almost a work of nature."

When completed, the Washington Monument was the tallest manmade structure of any kind in the world, but it held the title only briefly—until the wrought-iron Eiffel Tower was completed in 1889. However, the Washington Monument remains the world's tallest masonry structure to this day. The corrective measures taken on its foundation have proved to have been so well designed and executed that well over a century after the beginning of construction, the structure had settled only about four inches. The principal blemish of the monument is that it retains the ineradicable mark of the hiatus in its construction. About a quarter of the way up from the ground, there is a distinctly visible line of demarcation separating two colors of marble cladding the structure. Had the monument been constructed bottom to top at the same time, out of stone from the same quarry, any discoloration due to age and weathering would have been more or less uniform up the entire height of the structure. The situation when construction was resumed was not unlike trying to patch a section of a concrete sidewalk: getting the color of new concrete to match that of older is no easy task, the success of which is seldom achieved. Once a structure is completed, it is what it is.

I was once shaken out of my comfort zone not at the top of a monument or tall building in a big city but on the ground floor of a two-story motel in Klamath Falls, Oregon. On a beautiful spring morning in 1992, Catherine and I were in our room watching the news when an unfamiliar sensation overcame us. It felt as if the bed on whose edge we were sitting was swaying ever so gently, and the floor on which our feet rested seemed to be moving also. At the same time, we heard out in the hallway the noise of glasses rattling, which we figured came from a housekeeper's cart. Just before the floor and bed had begun to move, we had heard the housekeepers positioning their carts along the hallway in preparation for cleaning the rooms, and I thought that maybe the heavy carts being rolled along the hall had somehow caused our floor to shake and that motion had caused the bed to move. I realized differently only when the news was interrupted by a bulletin announcing that an earthquake had just occurred, with the epicenter in northern California. Indeed, the epicenter was near Eureka, California, about 150 miles from Klamath Falls.

Earthquakes are not necessarily accompanied by the rattle of glasses or the ringing of bells. Seismic forces might even be thought to be silent scourges, but they can be far from that to the very sensitive acoustic detectors that are used to provide prior warning even before seismographs can record any ground motion. Just as the wind can excite the natural frequencies and modes of vibration of structures, so can earthquakes. I have a favorite eighteen-inch-long plastic ruler that is quite flexible, and I have used it often to demonstrate to students what ground motion can do to a tall building. By grasping one end of the ruler, holding vertically and then moving my hand back and forth, I can show how differently the ruler responds to different frequencies and amplitudes of motion. If I move my hand slowly back and forth, the ruler moves with the hand and remains pretty much straight and vertical. If I shake my hand rapidly back and forth, the ruler responds by vibrating with a pronounced bending and whipping motion, with the precise shape depending on the frequency of the shaking. If I move my hand back and forth at the lowest natural frequency of the ruler, it whips back and forth so greatly that it looks as if the ruler might snap. While the plastic ruler has much more flexibility than a steel or concrete building, the same forces and responses to forces are at play, and it is the structural designer's challenge to keep the motion within limits so that occupants remain comfortable in the wind and the building stays together in an earthquake.

Another way an earthquake can do damage to a structure is by leaving it behind as the ground moves out from under it, the way a place setting is when a tablecloth is yanked out from beneath it. This happens with bridges, when their main girders rest upon but are not adequately tied down to their supports, and the phenomenon can be illustrated easily when the ends of a ruler are placed across the fingers of hands spaced almost as far apart as the ruler is long. If the thumbs are pressed down firmly on the ruler to hold it securely, then no matter how much or how fast the hands move back and forth in synchrony, the ruler will follow. If the thumbs are removed from the ruler, however, and the hands are moved rapidly back and forth, then the ruler will slide across the fingers and eventually drop off of them. It is the inertia of the ruler and the absence of the restraining forces of my thumbs that prevents it from keeping up with the motion of the fingers. In 1994, many highway bridges in southern California were destroyed by this mechanism during the Northridge earthquake. No structure, not even the stately Washington Monument, can withstand the forces of an earthquake if its parts are not properly secured or if its natural frequency is too close to that of the shaking ground.

The 2011 East Coast earthquake was the largest ever recorded that side of the Rocky Mountains. The Washington Monument as a whole moved with the ground motion, but because of the inertia of the individual stones there was also some relative movement among them. The resulting motion of stone on stone led to damage, opening up cracks in mortar between stones that could slide over one another and in some cases creating cracks in a single stone that through wedge or other action could not move freely. The crack in one stone near the top of the monument was more than an inch wide and allowed light and rain through the one-stone-thick inclined wall of the pyramidion. When the full extent of the damage to the structure was assessed by engineers, who rappelled down the sides of the faux obelisk, they identified other chipped and cracked stones and missing mortar. To enable repairs to proceed, scaffolding was erected all around the monument, just as it was in the late 1990s for the restoration work that was done in preparation for the Millennium. As he had been for that, the architect Michael Graves was commissioned to cover the scaffolding in a decorative scrim that bore the outline of stones of an exaggerated size. A lighting system installed behind the scrim illuminated the monument at night, making it a gigantic work of art.

With damaged stones repaired or replaced, Washington's grand needle stands today as a monument to hard work and perseverance. The gleaming obelisk on the hill is a reminder of how arduous building can be, raising an edifice piece by piece toward the sky. Looking at it and knowing its story, I can feel the effort of workers struggling with and nudging stones into place from its foundation to its stairway to its pyramidion. It is a feeling that I get also whenever I see a stone arch bridge or a dome or a cathedral. How the laborers making those edifices must have perspired as they felt the resistance of the stones to being moved into place! I think back to my youthful struggles with heavy bags of newspapers, feeling not only the arduousness of working to counter the forces they bore down with on my bicycle but also the forces I applied to flip them onto the porches along my paper route. I no longer wrestle with such efforts, but I can still feel the forces that I once experienced so viscerally. The objects I lift these days are lighter: books, silverware, pencils, and pens—and maybe a sleeping cat now and then—but the experience of heft continues to remind me of the role of forces in our daily lives.

20

Forces Felt and Heard

Precursors to an End

Inanimate objects, such as a block of stone or a length of lumber, may not feel forces in exactly the same way people do, but engineers freely speak about structures made of such components "feeling" and "reacting to" the loads they bear. It appears to be human nature to anthropomorphize just about everything nonhuman. No matter what cultural differences may prevail, we know or can imagine how it feels to be burdened with a large jug of water or to struggle under an awkward gathering of belongings balanced on our head or to bear on our shoulder a hod full of bricks that we carry up a ladder or to ferry on our back a weaker companion across a running stream. We may sympathize with the caryatids and attribute feelings to them, but do statues and monuments of any kind ever *really* feel the burden they support?

Does the Statue of Liberty feel in her arm the weight of the torch she holds in her hand? Do her iron muscles and tendons tire from the unrelieved pose that she maintains? These may seem to be impertinent or at best analogically inappropriate questions, for neither a caryatid nor Lady Liberty is a sentient being. That does not mean that statues do not feel heavy weights bearing down on their heads nor tire of rivets being stitched into their robes. Indeed, the laws of nature embodied in the natural mechanical principles uncovered by Newton determine the reactions of statues to the burdens that they carry as surely as the laws of anatomy and physiology tell us what we know by instinct: being pushed down upon by too heavy a load can be crushing, and holding too heavy a load above our head for too long can be tiring, regardless of whether we are a living person or an enduring statue.

Engineers are taught to think about the concrete in the abstract. For example, they learn to see a stone architrave spanning a pair of adjacent stone columns and a steel beam resting on a pair of steel columns as essentially the same structural arrangement—a horizontal element supported at its ends from beneath by a pair of vertical elements. The engineer's eye sees the beam as a geometric object subject to the forces of gravity (the beam's own weight plus whatever rests atop it) and the forces of connection and support (the upward push of the columns). The magnitude of the loads is measurable, and the reactions are calculable. The forces internal to the beam that hold it together (or, if they cannot, yield to its being broken apart) are also calculable. The downward deflection of the beam is governed by (and can be calculated using) Hooke's law, with its spring constant reflecting the stiffness of the material with which and in the configuration in which it was made.

But calculations, in engineering mechanics or any other science, are in themselves meaningless without an understanding of the starting assumptions upon which they are based and of a criterion under which they remain valid. Thus, in the case of the beam spanning the space between two columns, Hooke's law applies only if the beam is made of an elastic material—that is, one that allows the beam to return to its original shape when the load that bows it downward is removed. Furthermore, the results of calculations of internal stresses are meaningful only when compared to a failure criterion—that is, the limiting value for the material. It is when a critical value is reached that the beam breaks. This is what happened when stones in the Washington Monument cracked during the earthquake.

Even when the failure stress is not reached in a structural element, damage can be done. In the case of the arm of the Statue of Liberty, engineers found that the inadequately designed network of wrought iron struts and ties within the lady's limb had allowed stresses to come uncomfortably close to the breaking point. In calculating the stresses, to the effects of the dead weight of the arm and torch and their internal support frame had to be added that of the repeated loading and unloading from being buffeted by winds and heated and cooled through diurnal and seasonal cycles. Such fluctuating loads are associated with metal fatigue. As the statue's arm suffered the further repeated movement of crowds of people climbing through the truss-work into and out of the torch, it became clear that the tourist experience had to be curtailed and the ironwork reinforced. This was done in conjunction with the refurbishment that preceded Millennium celebrations.

People are not statues, of course, but they do have an underlying structure known as a skeleton. It is our skeleton, along with its interconnecting tendons and ligaments, that effectively provides support for our internal organs and tissues and our skin. It is also through our skeleton mainly that the burdens and loads we carry on our head and shoulders and back or by our arms and hands are transferred through our legs to the ground on which we stand and walk. If we try to overburden ourselves or are struck sharply, we can suffer cracked or broken bones, torn or ruptured tendons. Bones that break clean across can be thought of as beams that have been overloaded, as when a tackled football player's leg is fallen on hard by another player. Some bone fractures have a helical geometry, which can result when a leg that is planted solidly on the ground is twisted because the upper body turns suddenly to change direction in a way that the foot in its cleated or rubber-soled shoe cannot follow.

It is not only bones that break because of unusually violent impacts or movements. I once experienced a ruptured biceps tendon while rearranging furniture in preparation for visitors. I was attempting to move a day-bed out from a confined space, which meant that I had to exert pushing forces on the heavy bedframe with my arm fully extended. To accomplish this, I was asking the tendon stretched between my shoulder and elbow to pull beyond the limit of its connection—its breaking point. I heard what was to me an uncharacteristic "pop," but what I later learned from the internet and had confirmed by an orthopedic specialist was the characteristic sound of a tendon tearing away from the bone to which it is anchored. With the pop, I immediately lost all curling and turning ability in my arm—the motions needed to lift up a bag of groceries and turn a doorknob. In fact, I lost all strength and feel for performing even some of the simplest of tasks for daily life.

Surgery was not necessary or recommended for this proximal biceps tendon rupture, to use the medical terminology that classified my condition, but immobilization of the arm followed by physical therapy was. In time, I was told, other muscles and tendons would adjust to restore strength and motion to my arm. This was indeed the case, but it took about three months of wearing a sling before my arm's prior strength returned, and it took another month or so of physical therapy and gradually increasing what I asked the arm to do before I felt that it had returned to some semblance of normalcy. It was an experience in which I not only felt a force but also heard the part of my body applying it reach its limit. And to this day, I have a visual reminder of the incident through a lump in the crook of my arm indicating where the tendon remains bunched up.

The audible pop of a tearing tendon may be a one-time thing, but our bodies also make many repeated sounds, some voluntary and some involuntary. These can include cracking knuckles, creaking knees, and a snapping neck. All structures, in addition to the human skeleton, make noises as they move in the course of functioning and aging. Old houses are full of groans, squeaks, and creaks that usually grow more noticeable with age. Hardwood floors and stairs once securely held in place with fresh nails can begin to talk back after being trod upon over and over. The noise often comes from the hardwood flooring sliding across the subfloor or the subfloor across the joists. Noisy floors can also be the result of nails alternately sticking and slipping in the wood the way rubber-soled shoes do on a basketball court, the steel playing the floorboards and underlayment as if they were some kind of musical instrument. In time, the up-and-down motion of the boards can work the nails loose, which creates a palpable movement of the floor to accompany the sound.

In the first years we lived in our present house, I kept tripping on the heads of a couple of finishing nails that some previous owner must have unceremoniously driven into the hardwood to quell the feel and squeak of a particularly loose and loud floorboard. Now and then, I would pound the projecting nails down flush with the wood surface, but this fix was only temporary, for by the mechanics of stick and slip and ratcheting, the nails worked their way up again. Every spring, nails erupted from the floor the way weeds did in our lawn, but unlike Dylan Thomas's force that drove the flower, each nail seemed a steel fuse driving me mad. I finally used a claw hammer to extract the offensive flowers, stems, and taproots, and they yielded with a squeal that I heard with my ears and felt as a vibration in the handle of the tool. Without the nails, the floor was silent but spring-ier, something my family was happy to live with.

Over time, the house's roof developed leaks here and there in no dis-cernible geometric pattern. Roofers recommending a new roof reluctantly provided ad hoc fixes with patches of hot tar and new shingles that did not quite match the old. This did nothing to prevent leaks from devel-oping elsewhere, however. Finally, a roofer did sell us on the idea of an entirely new roof, based on his credible assessment that the problem was that the wood beneath the old shingles was rotting and so could not keep a firm grip on the roofing nails. Depending upon a complex interplay of forces having to do with heating and cooling, freezing and thawing, the broad-headed nails were working their way out of the underlayment, pop-ping up beneath shingles, pushing them away from the roof, and leaving an open drain of sorts for rainwater to get beneath them. His recommen-dation was to strip off all the old shingles and add a fresh layer of wood before shingling the roof anew. It seems to have worked; we have had no more leaks, but given the acoustic barrier known as an attic between our roof and living space, we are no more likely to hear nails popping now than we did before.

Even more so than wood-framed houses moved by the forces of wind and walking—as well as seasonal heating and cooling—boats and ships are prone to screeching and groaning as they float about in harbors and sail at sea. Rudyard Kipling famously captured the phenomenon in his short story "The Ship That Found Herself." On the cargo ship *Dimbula*'s maiden voyage, a Miss Frazier, the daughter of the owner, speaks to the Scottish captain about the vessel as a well-made structure. He informs her that a ship "is in no sense a reegid body closed at both ends. She's a highly com-plex structure o' various an' conflictin' strains, wi' tissues that must give an' tak' accordin' to her personal modulus of elasteecity." The ship's en-gineer joins the conversation and assures the young woman that a chris-tened ship is not a seaman's ship until she is "sweetened," and nothing but sailing through a gale will work out the kinks.

Soon after the woman was returned to Glasgow, the ship went down to Liverpool and took on four thousand tons of dead weight before setting sail for deep water. Kipling describes how the ballasted ship then became vocal: "As soon as she met the lift of the open water, she naturally began to talk. If you lay your ear to the side of the cabin, the next time you are in a steamer, you will hear hundreds of little voices in every direction, thrilling and buzzing, and whispering and popping, and gurgling and sobbing and squeaking exactly like a telephone in a thunder-storm. Wooden ships shriek and growl and grunt, but iron vessels throb and quiver through all their hundreds of ribs and thousands of rivets." The noises, the captain explains, are the various parts of the ship complaining about the weather and about their neighboring parts. The racket will continue until the ship finds herself, meaning the parts will have learned to work together. This happens to the *Dimbula* as she is entering New York Harbor. Indeed, it happens to every ship, for when she "finds herself all the talking of the separate parts ceases and melts into one voice, which is the soul of the ship."

Once, on a rough North Atlantic crossing, I heard the *Queen Elizabeth 2* speak. As I lay in my berth, I heard the steel hull flex and sing of journeys past. All structures do make sounds. Some of them are part of the normal sweetening process; others can be warnings of something abnormal. A typical main cable on a major suspension bridge may consist of tens of thousands of steel wires bundled together to form the long, heavy, outsize sagging clothesline that is slung over the towers and from which the roadway is hung. Especially when the wires must be protected from the corrosive effects of salt water and acid rain, the cables are wrapped in a steel sheathing and painted to keep pollutants at bay. While this action does generally prevent corrosion, chinks in the armor can develop, and the fact that the insides of the cable are covered up makes it difficult to inspect them for damage. There is a nonvisual way of monitoring the condition of the wires, however, and that is through a process known as acoustic emission sensing. It is well known that, just as a tearing tendon emits a pop, so a piece of steel that breaks emits an audible sound, a snap. By installing what are essentially small microphones on a bridge cable, interior wire failures can be detected and counted. When the number of failed wires reaches a dangerous percentage of the total making up the cable, critical engineering decisions have to be made.

The Waldo-Hancock suspension bridge in Maine was not being monitored for acoustic emissions, but when a portion of one of its cables was unwrapped in 1992 to check on the condition of the sixty-year-old structure, thirteen of the cable's more than thirteen hundred individual wires were found to be corroded and broken. Since less than 1 percent of the wires were affected, the bridge was allowed to continue to carry traffic but was scheduled for another inspection of its cables in ten years. When the cable was opened up in 2002, eighty-seven wires were found broken, comprising over 6 percent of the total. Engineers estimated that this lowered the bridge's factor of safety from 3 to about 2.4. It was clear that the structure's strength was deteriorating at an accelerating rate and something drastic had to be done. Heavy truck traffic was restricted on the bridge, and plans were made to replace the corroding structure with a new one.

Since the design and construction of a new bridge would take years, work was needed on the old crossing in the meantime, if it was to continue to carry traffic safely. The unusual measure taken was to add supplementary cables to the structure, thereby taking some of the load off the deteriorating ones. Now, even though the replacement bridge—the Penobscot Narrows Bridge and Observatory—is an innovative cable-stayed design, its steel components are threatened by the same corrosive saltwater environment as those of the old bridge were. To protect the new cables against deterioration, they are coated with epoxy and encased in high-density polyethylene pipes pressurized with nitrogen gas. The gas not only keeps out oxygen, without which corrosion cannot progress, but also allows the pressure in the pipes to be monitored constantly so that any leak that develops and threatens the steel cables can be immediately detected and repaired. It might be said that the cable system of the new bridge is designed to keep the steel wires from feeling much of anything but the load they are designed to carry. Furthermore, because of the innovative way in which the cables are installed, individual cable strands can be removed for inspection at any time, making an acoustic-emission system unnecessary. The bridge is designed to be seen and not heard.

The ideal condition in which a structural system might be said to exist would be one in which the forces that it feels are within its comfort range—that is, they are neither so large that there is danger of failure nor so small that the structure is not efficient. There is also the acoustic condition of structures. Just as every object has a natural frequency, so every structure has a natural ring to it. Baseball players demonstrate this when they tap their bat on home plate, expecting to hear a familiar sound. If instead they hear a dull thud, they conclude that their bat is cracked and so discard it for a sounder one. In a similar way, early railroad inspectors became adept at walking along the length of a stopped train tapping each of the wheels of the cars with a hammer. A discordant tone was a signal that the wheel was cracked and a more precise inspection or replacement was in order.

I have chosen in this book to focus mainly on the feel of forces and secondarily on the stiffness of structures. But stiffness is just a hammer's tap away from sound, and so I might just as easily have written first about the sound of structures and only incidentally about the forces involved. Yet choosing between strength and stiffness can seldom be done so arbitrarily in the real world of engineering. When engineers are charged with designing a bridge to span a river, they know first of all that the structure will have to be strong enough to withstand all the forces it might be subjected to, including its own dead weight, the weight of cars and trucks and pedestrians using the bridge, the rhythmic walking of parade goers or protesters or marching soldiers crossing the bridge, the weight of snow and ice that might accumulate on it in the winter, the force of the wind on it during a storm, and the shake of a sudden earthquake. At the same time, the bridge will have to be stiff enough that its roadway does not deflect excessively under any of those loads. A certain flexibility or bounciness to the roadway will be acceptable, but too much would not, as we have seen with the London Millennium Bridge. How much flexibility and movement is acceptable can be a matter of specification by the client commissioning the bridge, of regulation by governmental bodies overseeing such structures, of acceptance by the people using the bridge, or of just good judgment on the part of the engineers.

In some cases, the flexibility of a structure might in fact be considered the primary limitation, with sufficient stiffness following from a very strong design. This might be the case in the design of a laboratory building in which delicate instruments could be disturbed by too bouncy or too stiff a structure. In extreme cases, the instruments could be so affected as to give false readings during an experiment that demanded extreme sensitivity in the movements being measured and data being recorded. The usual remedy is to isolate laboratory spaces involving motion-causing experiments from those whose researchers cannot abide motion.

Some familiar structures might look and feel strong and stiff in isolation but behave differently in use. An aluminum beverage can is a structure classified by engineers as a pressure vessel. For it, the primary design consideration is to contain a carbonated beverage without leaking or exploding. A secondary consideration is to have on its top a device that also will not leak but will allow the can to be opened with ease. A further consideration is to make the can as light as possible, thereby using a minimum of material and thus making it relatively inexpensive to manufacture and distribute. Where the happy medium between strength and stiffness lies is a matter of judgment by the can manufacturing company, the bottling company, the retailer, and the consumer. Engineers designing the can structure should pay attention to all stakeholders; if their perceptions are incompatible, the engineers may be able to come up with some cleverly modified design that provides an acceptable compromise. One way of gaining some stiffness without thickening the wall or sacrificing strength is to introduce multiple small folds into the can's wall. This was tried by manufacturers some years ago, but the unorthodox multifaceted can was too easily dented and so did not survive in the marketplace, where appearance may matter above all else.

The design of a structure like an aluminum beverage can that at the same time appears to be remarkably simple but is in fact maddeningly complex will likely remain a challenge to engineers. Even when a seemingly satisfactory solution has been found, there will be subsequent challenges to it that use still thinner walls that result in less strong vessels that are even flimsier to the touch. The result may be exploding cans and ones that crush too easily in the hand. Such failures will make the new design unacceptable, but that will not stop inventors and creative designers from devising alternatives. One such could be thinner-walled, taller cans that maintain strength by their smaller diameter but sacrifice stiffness because of the longer distance between their top and bottom. The flexibility issue might be addressed by incorporating stiffening rings or flutes into the wall, but the new can would be such a departure from the familiar proportions and look of a common beverage can that consumers would likely reject it out of hand—if not because of the unfamiliar feel, then because of the unfamiliar sound that the can would make when popped open and set down.

The design of structures and systems will always involve a juggling of science and art. The engineering science of strength of materials enables an engineer to calculate the strength and stiffness of a structure, but it takes the good judgment of a keen eye and an astute ear for the engineer to see, feel, and hear when strength or stiffness has been relaxed too far, especially if one is given such importance over the other that what results is attractive neither to the hand, nor eye, nor ear. Developing at the same time a good feel for forces, a good eye for line, and a good ear for sound is the engineer's best way to avoid designing a weak, soft, clunker of a can or anything else.

Epilogue

A Forceful End

The bubonic epidemic known as the Great Plague broke out in London in 1665 and lasted for two years. During that time, citizens who could escape to the countryside did. Some fled to Cambridge, sixty miles to the north, where the university was shut down. This drove some of its residents to flee even farther from the epicenter.

Among those who did was twenty-three-year-old Isaac Newton. He had just finished his undergraduate studies and was teaching at his alma mater when the plague drove him to spend much of the next couple of years in and around his native hamlet of Woolsthorpe-by-Colsterworth, located about fifty miles northwest of Cambridge. He was in the prime of his intellectual life, and according to his own recollection, those years proved to be his most fruitful and creative. Indeed, according to the historian of science Robert Palter, it was during his sojourn in Lincolnshire that Newton "developed the integral calculus, experimentally verified the composite nature of light, and refined his gravitational theory to the point that he was able to satisfy himself through calculation that the earth's gravity holds the moon in orbit." Since most of this outburst of genius occurred in the single year 1666, scholars refer to it as Newton's *annus mirabilis*, his "miraculous year." Although there may be some hyperbole involved, Palter believes that the year "may surely be taken as symbolic of a decisive turning point in the history of human thought."

Similarly, the folksy anecdote about Newton being inspired to come up with his theory of gravity by an apple falling on his head may well be apocryphal, but that does not invalidate it as being universally evocative of an Aha! moment. It has been speculated that Newton did not actually get hit by an apple but merely observed one falling from a tree. Still, the idea that he did get physically bonked provides for an enticing vernacular alternative to technical explanations involving calculus, optics, and calculation—all terms appropriated effectively by politicians, by the way—of how he was inspired to formulate his theory of gravitation. Who cannot imagine a falling apple hitting them on the head? Who cannot imagine picking up the apple in anger and throwing it far from the tree? Who cannot see that all the while the apple is flying toward a compost heap it is falling back toward the Earth? Who cannot grant that this might inspire a vision of the Moon orbiting the Earth?

Many a person living long before Newton must have observed fruit falling from trees or rocks thrown at foes. They may also have experienced the physical consequences of being under a tree or on the wrong side in a feud at the wrong time. But observing and feeling do not necessarily lead to internalizing those sensations to the point of inspiration. Galileo came close to a broad understanding of gravity by connecting the phenomenon of a ball shot horizontally out of a cannon with that of one falling from a tower. Newton carried the image further by placing the cannon atop a high mountain and imagining the ball shot with such a speed that instead of eventually arcing back to strike the Earth it encircled it. Newton epitomized the value of imagery and metaphor to express universal concepts when in 1675 he wrote to Robert Hooke, "If I have seen further it is by standing on the shoulders of Giants," the statement often used to symbolize scientific progress. There is, however, an alternate interpretation of the oft-quoted words. Newton and Hooke were locked in debate over claims of priority when it came to gravitation. According to one view, when Newton referenced the shoulders of giants he was deliberately slighting Hooke, whose body was deformed and far from giantlike.

Especially in the area of mechanical force, progress does not demand abandonment of the past. We can empathize with Newton precisely because when we feel a force associated with everyday life today, we cannot imagine that it is different from what Newton and his colleagues felt it to be in the seventeenth century. And, by extension, they experienced force the same way their ancestors did. Thirteen centuries earlier, when Saint

Augustine and his friends shook a pear tree to steal its fruit, they must have felt the resistance of the tree's trunk to being bent in the same way that we can feel the stubbornness of a sapling when we might try to coax it to give up its last leaves in the fall.

The forces described in this book have always been part of the world. It may have taken some time for philosophers to articulate them and their role in effecting change, but the deepest thinkers were not necessarily the ones who first experienced or discovered these tangibles of the universe. That was done by cultures of individuals feeling in the course of everyday life how this thing we now call force in its myriad manifestations shaped our world and was somehow behind its sometimes puzzling behavior.

Artisans, craftspeople, inventors, and engineers never have had an absolute need for a theoretical basis to exploit the forces of nature. They have used them as they found them to be effective in context. Sentient beings learned to use a harder rock to knap a softer one: Anonymous workers in ancient Egypt did this literally to form everything from a cutting edge to an obelisk. Renaissance sculptors hit a harder chisel with a softer mallet to reshape chunks of stone into things of beauty: Michelangelo did so in sixteenth-century Florence to free his *David* from a block of marble. Modern problem solvers continue to use a hard pencil on soft paper (perhaps in the form of a computer and its output) to reshape the world: Engineers have done so in designing everything from bridges that span a mile to interplanetary probes that reach the outer planets and someday, maybe, spacecraft that carry astronauts there. It is human to learn by feeling and doing and thinking how to apply forces to achieve ends, no matter how intractable at first the problem may have seemed. Successive generations simply do on a grander scale what their ancestors had been doing for millennia and what children do today: they work and play with force. Perhaps it is because of their universality in time and space that such fundamental activities eventually become so unremarkable.

In the three and a half centuries since Newton's annus mirabilis, our everyday concept of force has served us well, and it can be expected to continue to do so. In spite of Einstein's theory of relativity explaining the gravitational force as a warping of a space-time continuum, and the ongoing quest of quantum mechanics for a theory of everything, the forces of Newton remain the mainstay of everyday life and terrestrial engineering, as well as of the flights of fancy that will reach their destination in some not yet fully shaped world.

The fact that Newton's annus mirabilis occurred during a period of plague should reassure us that in the long run the legacy of the Covid-19 pandemic need not include long-term adverse effects on artistic, scientific, and technological progress. The world's *anni miserabiles* are likely to be analyzed and reanalyzed intellectually for decades, if not centuries, even if many of our contemporary experiences are forgotten or become unimaginable to some of our descendants. But reflective twenty-second-century citizens should be able to replicate what it was like to wear a crudely designed mask, maintain an unsocial distance, and abstain from touching their hand to their face. And by feeling the past they should be better prepared to touch the future.

Acknowledgments

In the course of turning the germ of an idea for a book focused on how we feel force into a manuscript with a somewhat broader reach, I have relied on a wide variety of artifactual, digital, verbal, literary, video, and human resources. Many of the individuals from whom I benefitted through casual conversation, on the internet, and by mail—both snail and electronic—are so numerous that I dare not try to acknowledge them all. However, there are individuals whose help has been so selfless and inspiring that I do wish to mention them by name.

I am grateful to Professor Roland Paxton, of Heriot-Watt University, for his ongoing help with all things related to engineering in the United Kingdom and beyond. In response to a query regarding anthropomorphic models, he provided me with helpful information on the Forth Bridges Visitor Centre Trust and on the relocation of the human cantilever apparatus to South Queensferry, as well as the photo of the trustees posing on it. Michael Chrimes, former director of engineering policy and innovation at the Institution of Civil Engineers, London, took an interest in my search for a specific image of Michael Faraday lecturing. William Baker, structural engineering partner at Skidmore, Owings & Merrill, helped me locate an image representing in a human context the concept of buttressing employed in Burj Khalifa, for which he was the structural engineer. Maria Garlock, professor of civil and environmental engineering at Princeton University, along with Robert Reitherman, executive director of the Consortium of Universities for Research in Earthquake Engineering, helped me obtain the image showing models of the two truss designs of the Golden Gate Bridge. Arup's Daniel Imade kindly provided me a selection of images of the London Millennium Bridge from which to choose. Braulio Agnese of the National Building Museum graciously offered photos of arch building in the Great Hall. William Walters and Amy Surak of the Manhattan College Library tracked down the given names of Mr. Zia.

I am also grateful to Professor Chris Calladine for inviting me to visit Cambridge University, where he introduced me to Professor Sergio Pellegrino and his Deployable Structures Laboratory. I benefitted greatly from guided tours of their laboratories and from our most interesting discussions about tape measures and deployable structures generally. Consultant engineer James F. Edwards, whose article on construction methods used in building the Great Pyramid at Giza led to our corresponding on the subject, graciously elaborated upon his fresh ideas. Nick Hill facilitated my communication with Dr. Edwards, who does not have a computer. To the anonymous reader who identified some shortcomings in the manuscript for this book, I am indebted for pointing out potentially misleading statements about the effectiveness of surgical masks and the nature of the relationship between Isaac Newton and Robert Hooke, which I had the opportunity to clarify before the book went to press.

I am grateful to readers of *American Scientist* who provide helpful and insightful feedback on my regular "Engineering" column in the magazine, as well as suggestions for future topics. Some of the material in this book first appeared, usually in considerably different form, in *American Scientist*. The magazine *Metropolis* was where I first expressed some of my ideas about the book as an uncomfortable artifact. I first explored some ideas related to indoor transportation in an essay in the catalog *Up Down Across: Elevators, Escalators, and Moving Sidewalks* accompanying the eponymous exhibit at the National Building Museum. These essays are referenced at their appropriate place in the Bibliography.

I am grateful to Duke University in general, and to its Pratt School of Engineering in particular, for having provided a nurturing environment for me to explore so many different aspects of engineering and technology over my forty-year association with the institution. Its library has been superb for my purposes, and its librarians have been invaluable and patient resources without whom I could not have pursued such extensive research on some of my more unconventional writing projects. I am grateful to the Duke and Pratt administrators who most recently supported my request for a sabbatical leave to develop an early draft of this book.

Andrew Stuart, my agent for this book, saw its possibilities in an early draft of a proposal and encouraged me to be more ambitious in my execution. I am glad I followed his advice, and I am grateful to him for making possible my association with Yale University Press. At the press, I am indebted to my editor, Jean Thomson Black, whose fondness and enthusiasm for the project have remained strong throughout the acquisition, production, and publication process. Indeed, I incorporated into the final draft some of the anecdotes she related during our animated telephone conversations about the book and its subject matter. Throughout the editorial process, her assistants Elizabeth Sylvia and Amanda Gerstenfeld were always responsive to inquiries and forthcoming with help. Joyce Ippolito, the production editor, did a marvelous job of overseeing the process of taking the book from manuscript to printed and bound artifact. Laura Jones Dooley was a sensitive and meticulous copyeditor. It was a pleasure to work with everyone who had a mind and hand in making this book what it is.

I might not have been able to work on, let alone complete, this book had it not been for my oncologist, Michael R. Harrison, M.D., of Duke University Medical Center. It was he who identified a research study for which I qualified genetically and advised me to enroll in it. Were it not for that action and the medication it provided me, my condition would almost certainly have continued to deteriorate and made it increasingly difficult for me to write. I am deeply indebted to Dr. Harrison and the sponsors of the project. I am also grateful to Kelly Onyenwoke, R.N., research project coordinator, who was enormously helpful throughout the study.

And, as I have been with all of my books, I am unreservedly indebted to my wife, Catherine, for being my first reader and my fairest critic. Her editorial assistance and proofreading have once again saved me from embarrassing faux pas. In this case, she has also been my prime photographer, videographer, and graphic services expert in preparing digitized images for the figures herein. And once again she has provided a flattering photo of me. As Catherine and I move through life together, the forces of attraction that drew us to each other in the first place continue to grow ever stronger.

Bibliography

Prologue

Bedient, Calvin. "Wanted: An Original Relation to the Universe." *New York Times,* December 22, 1974.

Earnest, Mark. "On Becoming a Plague Doctor." *New England Journal of Medicine* 383 (2020): e64. https://doi.org/10.1056/nejmp2011418.

Emerson, Ralph Waldo. *Nature.* Boston: James Munroe, 1836.

Fletcher, Robert. *A Tragedy of the Great Plague of Milan in 1630.* Baltimore: Lord Baltimore Press, 1898.

Halstead, John. "The Transcendentalists: An Original Relation to the Universe." *Naturalistic Paganism,* July 21, 2015. https://naturalisticpaganism.org/2015/07/21/the-transcendentalists-an-original-relation-to-the-universe-by-john-halstead.

Lohner, Svenja. "Test the Strength of Hair: A Hairy Science Project from Science Buddies." *Scientific American,* November 28, 2019. www.scientificamerican.com/article/test-the-strength-of-hair.

Matuschek, Christiane, et al. "The History and Value of Face Masks." *European Journal of Medical Research* 25 (2020): 23. https://doi.org/10.1186/s40001-020-00423-4.

Mussap, Christian J. "The Plague Doctor of Venice." *Internal Medicine Journal* 49, no. 5 (2019): 671–76.

Palter, Robert, ed. *The Annus Mirabilis of Sir Isaac Newton.* Cambridge, Mass.: MIT Press, 1971.

Sitter, John E. "About Ammons' 'Sphere.'" *Massachusetts Review* 19, no. 1 (1978): 201–12.

Strasser, B. J., and Thomas Schlich. "A History of the Medical Mask and the Rise of Throwaway Culture." *Lancet* 396, no. 10243 (2020): 19–20.

Thilmany, Jean. "High Standards: As Elevators Advance, So Do Their Safety Codes." *Mechanical Engineering,* July 2013, 46–49.

Thoreau, Henry David. *Walden and Civil Disobedience*. New York: Barnes and
   Noble, 2005.
Vence, Tracy. "Here to Help: How to Defog Glasses When Wearing a Mask." *New
   York Times,* March 25, 2021.
Wikipedia. "Plague Doctor." https://en.wikipedia.org/wiki/Plague_doctor.
———. "Plague Doctor Costume." https://en.wikipedia.org/wiki/Plague_doctor
   _costume.
———. "Surgical Mask." https://en.wikipedia.org/wiki/Surgical_mask.

1. Pushes and Pulls

Faraday, Michael. *A Course of Six Lectures on the Chemical History of a Candle*.
   Edited by William Crookes. London: Griffin, Bohn, 1861.
———. *A Course of Six Lectures on the Various Forces of Matter and Their Relations
   to Each Other*. Edited by William Crookes. London: Richard Griffin, 1860.
Galileo. *Dialogues Concerning Two New Sciences*. Translated by Henry Crew and
   Alfonso de Salvio. New York: Dover, [1954].
Hammack, William S., and Donald J. DeCoste. *Michael Faraday's "The Chemical
   History of a Candle": With Guides to Lectures, Teaching Guides and Student
   Activities*. Urbana, Ill.: Articulate Noise Books, 2016.
Huxley, Thomas Henry. "On a Piece of Chalk." *Macmillan's Magazine*, 1868.
   Reprinted as *On a Piece of Chalk*. New York: Scribner, 1967.
Jammer, Max. *Concepts of Force: A Study in the Foundations of Dynamics*. New York:
   Harper Torchbooks, 1962.
Krulwich, Robert. "Thinking Too Much About Chalk." *NPR*, July 12, 2012. www
   .npr.org/sections/krulwich/2012/07/12/156629934/thinking-too-much-about
   -chalk.
Royal Institution. "Complete List of Christmas Lectures." *Royal Institution of
   Great Britain,* undated. www.rigb.org/docs/christmas_lecturers_18252015_0
   .pdf.
Wolfle, Dael. "Huxley's Classic of Explanation." *Science* 156, no. 3776 (1967):
   815–16.

2. Gravitation

Babson, Roger W. *Actions and Reactions: An Autobiography of Roger W. Babson*.
   New York: Harper and Brothers, 1935.
"Babson's Boulders: A Millionaire's Odd Dogtown Legacy." *New England Histori-
   cal Society,* last updated 2021. www.newenglandhistoricalsociety.com/babsons
   -boulders-millionaires-odd-dogtown-legacy.
Bernstein, William J. *Masters of the Word: How Media Shaped History*. New York:
   Grove Press, 2013.
Corn, Alfred. "Gravitational." *New Yorker,* November 30, 2020, 54–55.
Dunavin, Davis. "The Man Who Defied Gravity." *WHSU Public Radio,* March 20,
   2020. www.wshu.org/post/man-who-defied-gravity#stream/0.

Emory University. "History and Traditions: Gravity Monument." Undated.
   https://emoryhistory.emory.edu/facts-figures/places/landmarks/gravity.html.
Gardner, Martin. *Fads and Fallacies in the Name of Science.* New York: Dover, 1957.
Gravity Research Foundation. "Founding of the Gravity Research Foundation."
   Undated. www.gravityresearchfoundation.org/historic.
———. "2021 Awards for Essays on Gravitation." Undated. www.gravityresearch
   foundation.org/competition.
Hopkins, Pamela. "A Grave for Gravity: How Tufts Pranksters 'Helped' with Anti-
   Gravity Research." Tufts University, Digital Collections and Archives, November
   18, 2016. http://sites.tufts.edu/dca/2016/11/18/a-grave-for-gravity-how-tufts
   -pranksters-helped-with-anti-gravity-research/.
Jammer, Max. *Concepts of Force: A Study in the Foundations of Dynamics.* New York:
   Harper Torchbooks, 1962.
Johnson, George. "Still Exerting a Hold on Science." *New York Times,* June 24, 2014.
Kaiser, Dave, and Dean Rickles. "The Price of Gravity: Private Patronage and the
   Transformation of Gravitational Physics After World War II." *Historical Studies in
   the Natural Sciences* 48, no. 3 (2018): 338–79.
Kaku, Michio. *The God Equation: The Quest for a Theory of Everything.* New York:
   Doubleday, 2021.
McCloskey, Michael, Alfonso Caramazza, and Bert Green. "Curvilinear Motion in
   the Absence of External Forces: Naive Beliefs About the Motion of Objects."
   Science 210, no. 4474 (1980): 1139–41.
M. J. L. "Defying Gravity." *Emory Magazine* 77, no. 3 (2001). www.emory.edu
   /EMORY_MAGAZINE/autumn2001/enigma.html.
Newman, James R., ed. *The World of Mathematics.* Vol. 3. Mineola, N.Y.: Dover,
   2000.
"Professor Faraday Lecturing at Royal Institution," *Calisphere.* https://calisphere
   .org/item/8b3766e6980b2930b4537d17becc4d85.
Pynchon, Thomas. *Gravity's Rainbow.* New York: Viking, 1973.
Thompson, D'Arcy Wentworth. *On Growth and Form.* Rev. ed. New York: Dover,
   1992.
Thomson, James A. "Beyond Superficialities: Crown Immunity and Constitutional
   Law." *Western Australia Law Review* 20, no. 3 (1990): 710–25.
"What the Heck Is a Hammerschlagen?" Darien Cornfest (website), undated.
   http://dariencornfest.us/?p=270.
Wolchover, Natalie. "How One Man Waged War Against Gravity." *Popular Science,*
   March 15, 2011. www.popsci.com/science/article/2011-03/gravitys-sworn-enemy
   -roger-babson-and-gravity-research-foundation.

3. Magnetism

Brake, Walter J. "Magnetic Novelty." U.S. Patent No. 2,249,454 (1941).
Norton, Quinn. "A Sixth Sense for a Wired World." *Wired,* July 6, 2006. www
   .wired.com/2006/06/a-sixth-sense-for-a-wired-world/.

Sklar, Mikey. "A Reasonably Priced Sixth Sense." *Holy Scrap* (blog), December 30, 2011. http://blog.holyscraphotsprings.com/2011/12/reasonably-priced-sixth-sense.html.

Taub, Eric A. "Hyperloop Technology Reaches a Milestone with Passenger Test." *New York Times,* November 10, 2020.

4. Friction

Acharya, R., et al. "The Ultrafast Snap of a Finger Is Mediated by Skin Friction." *Journal of the Royal Society Interface,* November 17, 2021. doi: 10.1098/rsif.2021.0672.

Azadeh, Payam. "Fingerprint Changes Among Cancer Patients Treated with Paclitaxel." *Journal of Cancer Research and Clinical Oncology* 143, no. 4 (2017): 693–701.

Belluck, Pam. "Cat Shows Survivor Instincts in 200-Mile Journey Home." *New York Times,* January 20, 2013.

Branch, John. "The Science Behind the Squeak." *New York Times,* March 18, 2017.

Bubola, Emma. "After Slips and Falls, Venice Gets a Grip on a Star Architect's Bridge." *New York Times,* January 3, 2022.

Galileo. *Dialogues Concerning Two New Sciences.* Translated by Henry Crew and Alfonso de Salvio. New York: Dover, [1954].

Gladwell, Malcolm. "Complexity and the Ten-Thousand-Hour Rule." *New Yorker,* August 21, 2013.

———. *Outliers: The Story of Success.* New York: Little, Brown, 2008.

Harmon, Christy. "First Weave a Classic Gag, Then Find a Victim." *New York Times,* January 24, 2021.

Hertz, Heinrich. *The Principles of Mechanics Presented in a New Form.* Translated by D. E. Jones and J. T. Walley. New York: Dover, 1956.

Walker, Alissa. "Venice Bridge Will Be De-Calatrava'd to Keep Pedestrians from Face-planting." *New York: Curbed,* January 2022. www.curbed.com/2022/01/venice-calatrava-slippery-bridge-fixed.html.

5. Fractious Forces

Fitzpatrick, Tony, and Roger Ridsdill Smith. "Stabilising the London Millennium Bridge." *Ingenia Online,* August 2001. www.ingenia.org.uk/Ingenia/Articles/3340e992-d4c5-485f-996b-e3c827856452.

Foderaro, Lisa W. "A New Bridge Bounces Too Far and Is Closed Until the Spring." *New York Times,* October 4, 2014.

———. "Repairs Coming for Bridge That Bounced Too Much." *New York Times,* July 14, 2016.

———. "Rhythmically but Safely: That's the Way the Bridge Bounces." *New York Times,* April 18, 2017.

Galileo. *Dialogues Concerning Two New Sciences.* Translated by Henry Crew and Alfonso de Salvio. New York: Dover, [1954].

McCann, Michael J. "Engineered for Trust: American Scientist." Email message to the author, June 14, 2021.

McCullough, David. *The Great Bridge*. New York: Simon and Schuster, 1972.

Ouellette, Jennifer. "New Study Challenges Popular Explanation for London's Infamous 'Wobbly Bridge.'" *Arstechnica*, December 20, 2021. doi: 10.1038 /s41467-021-27568-y.

Petroski, Henry. "Problematic Pedestrian Bridges." *American Scientist*, November–December 2017, 340–43.

Sheikh Zayed Grand Mosque Center. *Spaces of Light (Season 2): Sheikh Zayed Grand Mosque in Photographs*. Abu Dhabi: Sheikh Zayed Grand Mosque Center, 2012.

Sudjic, Deyan. *Blade of Light: The Story of London's Millennium Bridge*. London: Penguin Press, 2001.

6. Lever, Lever, Cantilever

Aristotle. *Minor Works*. Translated by W. S. Hett. Cambridge, Mass.: Harvard University Press, 1980.

Bloomfield, Samuel. "Can Opener." U.S. Patent No. 2,412,946 (1946).

Coates, John F. "The Trireme Sails Again." *Scientific American* 260, no. 4 (1989): 96–105.

Darqué, Etienne Marcel. "Tin-Box Opener." U.S. Patent No. 1,082,800 (1913).

Elson, Henry W. *Modern Times: And the Living Past*. New York: American Book, 1921.

Epstein, Marcelo, and Walter Herzog. "Aspects of Skeletal Muscle Modelling." *Philosophical Transactions of the Royal Society of London B* 358 (2003): 1445–52.

Foster, Renita. "The Greatest Army Invention Ever." *Pentagram*, August 18, 1986, 11.

Galileo. *Dialogues Concerning Two New Sciences*. Translated by Henry Crew and Alfonso de Salvio. New York: Dover, [1954].

Gugliotta, Guy. "The Ancient Mechanics and How They Thought." *New York Times*, April 1, 2005.

Hirschfeld, N. E. "Appendix G: Trireme Warfare in Thucydides." In *The Landmark Thucydides: A Comprehensive Guide to the Peloponnesian War*. Edited by R. B. Strassler. New York: Free Press, 1996.

Petroski, Henry. "Bottle and Can Openers as Levers." *American Scientist*, March–April 2017, 90–93.

———. "The Cantilever." *American Scientist*, September–October 2007, 394–97.

———. *Invention by Design: How Engineers Get from Thought to Thing*. Chapter 5. Cambridge, Mass.: Harvard University Press, 1996.

———. *Small Things Considered: Why There Is No Perfect Design*. Pages 168–71. New York: Alfred A. Knopf, 2003.

———. "Uncomfortable but Comforting." *Metropolis Magazine*, April 2007, 60, 62.

Salvadori, Mario. *Why Buildings Stand Up: The Strength of Architecture*. New York: McGraw-Hill, 1982.

Sampson, D. F., and J. M. Hothersall, "Container Opener." U.S. Patent No. 1,996,550 (1935).

Schlitz Beer. "Some Day All Beer Cans Will Open This Easy!" Advertisement. *Playboy*, 1962.

Speaker, John W. "Pocket Type Can Opener." U.S. Patent No. 2,413,528 (1946).

Strengberg, Dewey M. "Can Opener." U.S. Patent No. 1,669,311 (1928).

Szalay, Alexander. "Lever Adapter for Door Knobs." U.S. Patent No. 4,783,883 (1988).

Weiner, Eric. *The Geography of Genius: Lessons from the World's Most Creative Places.* New York: Simon and Schuster, 2016.

Witz, Billy. "How Judge Built a Mighty Swing." *New York Times*, July 17, 2017.

7. Forces, Forces Everywhere

Grecco, Pasquale. "Retrofit Lever Handle Used by a Disabled Person for Turning a Door Knob." U.S. Patent No. 4,971,375 (1990).

Greve, Frank. "Doorknob's Days Are Numbered." *Herald-Sun* (Durham, N.C.), December 27, 2007.

Jones, Robert L., Jr. "Lever Action Retrofit Door Handle." U.S. Patent No. 5,231,731 (1993).

Leopoldi, Norbert. "Lever Adapter for Door Knobs." U.S. Patent No. 4,877,277 (1989).

Perry, Eugene H. "Lever Door Handle." U.S. Patent No. 4,502,719 (1985).

Petroski, Henry. "Opening Doors." *American Scientist*, March–April 2012, 112–15.

United States Access Board. "ADA Accessibility Guidelines for Buildings and Facilities." Section 4.13.9. 2002. www.access-board.gov/adaag-1991-2002.html.

Walls, Laura Dassow. *Henry David Thoreau: A Life.* Chicago: University of Chicago Press, 2017.

8. Moments of Inertia

Chokshi, Niraj. "Why Cargo Ships Grew So Big." *New York Times*, March 31, 2021.

Flegenheimer, Matt. "In Expansion of No. 7 Line, One Problem: An Elevator." *New York Times*, May 30, 2014.

Goetz, Alisa, ed. *Up Down Across: Elevators, Escalators, and Moving Sidewalks.* London and Washington, D.C.: Merrell and National Building Museum, 2003.

Goldstein, Richard. "Fighter Ace and Test Pilot Embodied 'the Right Stuff.'" *New York Times*, December 9, 2020.

Goodman, Peter S. "Pileup at Port Is Becoming a Quagmire." *New York Times*, October 11, 2021.

Harris, Elizabeth A. "Supply Issues Are Causing Book Delays." *New York Times*, October 5, 2021.

Hertz, Heinrich. *The Principles of Mechanics Presented in a New Form.* Translated by D. E. Jones and J. T. Walley. New York: Dover, 1956.

Kareklas, Kyriacos, Daniel Nettle, and Tom V. Smulders. "Water-Induced Finger Wrinkles Improve Handling of Wet Objects." *Biology Letters* 9 (2013): 2.

Levinson, Marc. *The Box: How the Shipping Container Made the World Smaller and the World Economy Bigger.* Princeton, N.J.: Princeton University Press, 2006.

Penney, Veronica. "How Coronavirus Has Changed New York City Transit Ridership." *New York Times,* March 10, 2021.

Petroski, Henry. *Paperboy: Confessions of a Future Engineer.* New York: Alfred A. Knopf, 2002.

Swanson, Ana, Joanna Smialek, and Jim Tankersley. "Biden, Fighting Supply Chain Woes, Announces Port Will Operate 24/7." *New York Times,* October 14, 2021.

"A Tumbling T-Handle in Space: The Dzhanibekov Effect." *Rotations* (blog), March 25, 2021. https://rotations.berkeley.edu/a-tumbling-t-handle-in-space.

Veritasium. "The Bizarre Behavior of Rotating Bodies, Explained." Video, September 19, 2019. www.youtube.com/watch?v=1VPfZ_XzisU&feature=youtu.be.

Wikipedia. "Tennis Racket Theorem." https://en.wikipedia.org/wiki/Tennis_racket_theorem.

Yee, Vivian. "Suez Canal Bottleneck Disrupts Global Deliveries, Prompting Syria to Ration Fuel." *New York Times,* March 29, 2021.

Yee, Vivian, and James Glanz. "How a Massive Ship Jammed the Suez Canal." *New York Times,* July 18, 2021, 8–9.

Yee, Vivian, and Peter S. Goodman. "Suez Canal Blocked After Giant Container Ship Gets Stuck." *New York Times,* March 25, 2021.

9. Forceful Illusions

Bernoulli, Daniel. *Hydrodynamics.* Translated by T. Carmody and H. Kobus. New York: Dover, 1968.

"The Challenger Disaster." *RichardFeynman.com* (blog), undated. www.feynman.com/science/the-challenger-disaster.

Faraday, Michael. "On a Piece of Chalk." *Macmillan's Magazine* 18 (1868): 396–408.

Feltman, Rachel. "A Famous Physicist's Simple Experiment Showed the Inevitability of the Challenger Disaster." *Washington Post,* January 27, 2016.

Hertz, Heinrich. *The Principles of Mechanics Presented in a New Form.* Translated by D. E. Jones and J. T. Walley. New York: Dover, 1956.

Hewitt, Paul G. "The Pulled Spool—Which Way Does It Roll?" *Science Teacher,* April/May 2020. www.youtube.com/watch?v=DkkqhzatasI.

John Wiley and Sons. Promotional flyer for *Engineering Mechanics, Third Edition,* by J. L. Meriam and L. G. Kraige. [1992.]

Klinger Scientific Apparatus Corp. *Experiments in Optics, Part 2.* Bulletin 101-2. Jamaica, N.Y.: Klinger, [1963].

Lohner, Svenja. "Test the Strength of Hair: A Hairy Science Project from Science Buddies." *Scientific American,* November 28, 2019. www.scientificamerican.com/article/test-the-strength-of-hair.

Meriam, J. L., and L. G. Kraige. *Engineering Mechanics.* 3rd ed. New York: Wiley, 1992.

Palmer, A. N. *Palmer's Guide to Business Writing.* Cedar Rapids, Iowa: Western Penmanship, 1894.

Petroski, Henry. "Impossible Points, Erroneous Walks." *American Scientist,* March–April 2014, 102–5.

———. *The Pencil: A History of Design and Circumstance.* New York: Alfred A. Knopf, 1990.

———. *To Engineer Is Human: The Role of Failure in Successful Design.* New York: St. Martin's Press, 1985.

———. "Work and Play." *American Scientist,* May–June 1999, 208–12.

Rothman, Joshua. "Jambusters." *New Yorker,* February 12 and 19, 2018, 42–46.

Stoffer, Jim. "The Mail: All Jammed Up." Letter. *New Yorker,* February 26, 2018.

"Tom Brady Suspension Case Timeline." *NFL.com,* July 15, 2016. www.nfl.com /news/tom-brady-suspension-case-timeline-0ap3000000492189.

Weingardt, Richard G. *Circles in the Sky: The Life and Times of George Ferris.* Reston, Va.: ASCE Press, 2009.

Whewell, William. *An Elementary Treatise on Mechanics.* Cambridge: J. Deighton & Sons, 1819.

Wikipedia. "Royal Institution Christmas Lectures." https://en.wikipedia.org/wiki /Royal_Institution_Christmas_Lectures.

10. From Physics to the Physical

Bialik, Carl. "There Could Be No Google Without Edward Kasner." *Wall Street Journal Online,* November 30, 2016. www.wsj.com/articles/SB10857592492 1724042.

Holmes, Oliver Wendell. "The Deacon's Masterpiece, or the Wonderful 'One-Hoss-Shay': A Logical Story." *Atlantic Monthly,* September 1858.

———. *The Wonderful "One-Hoss-Shay" and Other Poems.* New York: Frederick A. Stokes, 1897.

Kasner, Edward, and James R. Newman. *Mathematics and the Imagination.* New York: Simon and Schuster, 1940.

Laplace, Pierre-Simon. *A Philosophical Essay on Probabilities.* Translated by F. W. Truscott and F. L. Emory. New York: Dover, 1951.

McCloskey, Michael, Alfonso Caramazza, and Bert Green. "Curvilinear Motion in the Absence of External Forces: Naive Beliefs About the Motion of Objects." *Science* 210, no. 4474 (1980): 1139–41.

Thompson, D'Arcy Wentworth. *On Growth and Form.* Rev. ed. New York: Dover, 1992.

Thomson, James A. "Beyond Superficialities: Crown Immunity and Constitutional Law." *Western Australia Law Review* 20, no. 3 (1990): 710–25.

11. Forces on Inclined Planes

Ames, Nathan. "Revolving Stairs." U.S. Patent No. 25,076 (1859).

Baker, Nicholson. *The Mezzanine: A Novel*. New York: Weidenfeld and Nicolson, 1988.

Goetz, Alisa, ed. *Up Down Across: Elevators, Escalators, and Moving Sidewalks*. London and Washington, D.C.: Merrell and National Building Museum, 2003.

Hitchens, Derek. "Pyramids Have Different Slopes." *Prof's Ancient Egypt*, undated. https://egypt.hitchins.net/pyramid-myths/pyramids-have-different.html.

"JFK Jr. Killed in Plane Crash." *History Channel*, undated. www.history.com/this -day-in-history/jfk-jr-killed-in-plane-crash.

Leonard, Eric. "Cause of Kobe Bryant's Helicopter Crash Finalized in NTSB Report." *NBC Los Angeles*, February 25, 2021. www.nbclosangeles.com/news /national-international/kobe-bryant-helicopter-crash-cause-finalized-ntsb-report /2536402.

Reno, Jesse W. "Endless Conveyer or Elevator." U.S. Patent No. 470,918 (1892).

Wikipedia. "Great Pyramid of Giza."

12. Stretching and Squeezing

Baltzley, Louis E. "Paper-Binding Clip." U.S. Patent No. 1,139,627 (1915).

Bittman, Mark. "Eat: Slawless." *New York Times Magazine*, March 17, 2013, 46–47.

Freeman, Mike. "Clarence Saunders: The Piggly Wiggly Man." *Tennessee Historical Quarterly* 51, no. 3 (1992): 161–69.

Hales, Linda. "A Big Clip Job? Think Washington." *Washington Post*, May 20, 2006.

Hamblin, Dora Jane. "What a Spectacle! Eyeglasses, and How They Evolved." *Smithsonian*, March 1983, 100–112.

Hooke, Robert. *Lectiones Cutlerianae; or, A Collection of Lectures: Physical, Mechanical, Geographical, and Astronomical*. London: Royal Society, 1679.

Ilardi, Vincent. "Eyeglasses and Concave Lenses in Fifteenth-Century Florence and Milan: New Documents." *Renaissance Quarterly* 29, no. 3 (1976): 341–60.

Jardine, Lisa. *The Curious Life of Robert Hooke: The Man Who Measured London*. New York: HarperCollins, 2004.

Letocha, Charles E. "The Origin of Spectacles." *Survey of Ophthalmology* 31, no. 3 (1986): 185–88.

Petroski, Henry. "The Evolution of Eyeglasses." *American Scientist*, September–October 2013, 334–37.

———. "Shopping by Design." *American Scientist*, November–December 2005, 491–95.

———. *Small Things Considered: Why There Is No Perfect Design*. New York: Alfred A. Knopf, 2003.

Rosen, Edward. "The Invention of Eyeglasses." *Journal of the History of Medicine and Allied Sciences* 11 (1956): 13–46, 183–218.

Rubin, M. L. "Spectacles: Past, Present, and Future." *Survey of Ophthalmology* 30, no. 5 (1986): 321–27.

Rucker, C. Wilbur. "The Invention of Eyeglasses." *Proceedings of the Staff Meetings of the Mayo Clinic* 35, no. 9 (1960): 209–16.

Saunders, Clarence. "Self-Serving Store." U.S. Patent No. 1,242,872 (1917).

13. A Round Cake in a Square Box

Beck, Dilman A., and Susan E. Beck. "Combination Food Server and Container Lid Support." U.S. Patent No. 4,877,609 (1989).

Coomes, Steve. "Thinking Round the Box." *Pizza Marketplace,* March 23, 2005. www.pizzamarketplace.com/news/thinking-round-the-box/.

Fisk, James, Jr. "Pizza Box with Wedgeshaped Break-down Spatula-plates." U.S. Patent No. 5,476,214 (1995).

Grynbaum, Michael M. "A Fork? De Blasio Is Mocked for the Way He Eats Pizza." *New York Times,* January 11, 2014.

Levinson, Marc. *The Box: How the Shipping Container Made the World Smaller and the World Economy Bigger.* Princeton, N.J.: Princeton University Press, 2006.

Lukas, Paul. "A Large Pepperoni, and Don't Skimp on the Cheese." *Uni Watch,* February 1, 2013. www.uni-watch.com/2013/02/01/a-close-look-at-the-little -doohickey-in-the-center-of-a-delivery-pizza.

Maultasch, Jonathan, and Bruce Maultasch. "Combined Pizza Box Lid Support and Cutter." U.S. Patent No. 5,480,031 (1996).

Nelson, David C., and John J. Andrisin. "Pizza Box Lid Support and Serving Aid." U.S. Patent No. 7,191,902 (2007).

Petroski, Henry. "A Round Pie in a Square Box." *American Scientist,* July–August 2011, 288–92.

———. *Small Things Considered: Why There Is No Perfect Design.* New York: Alfred A. Knopf, 2003.

Ramzy, Austin. "Olympic Beds Are Cardboard, Yes. But Sturdy. Just Trust Us." *New York Times,* July 20, 2021.

Ronan, Alex. "A Pile of Scrap Cardboard Inspired Frank Gehry's Iconic Collection." *Dwell,* March 19, 2015. www.dwell.com/article/a-pile-of-scrap-cardboard-inspired -frank-gehrys-iconic-collection-947ebbao.

Vitale, Carmela. "Package Saver." U.S. Patent No. 4,498,586 (1985).

Voves, Mark A. "Pizza Cutting and Eating Tool." U.S. Patent No. Des. 425,376 (2000).

Walsh, Savannah. "Watch Diana Meet the Queen in *The Crown* Season 4 Trailer: 'If She Doesn't Bend, She Will Break.'" *Elle,* October 29, 2020. www.elle.com /culture/movies-tv/a34519836/the-crown-season-4-trailer-diana-meets-the -queen.

Wiener, Scott. *Viva La Pizza! The Art of the Pizza Box.* Brooklyn, N.Y.: Melville House, 2013.

Wikipedia. "Pizza Saver." https://en.wikipedia.org/wiki/Pizza_saver.

14. Deployable Structures

Boroughs, Don. "Folding Frontier." *ASEE Prism,* January 2013, 24–29.

Brown, William G. "Coilable Metal Rule." U.S. Patent No. 3,121,957 (1964).

Calladine, C. R. "The Theory of Thin Shell Structures, 1888–1988." *Proceedings of the Institution of Mechanical Engineers* 202, A3 (1988): 141–49.

Li, Shih Lin. "Tape Rule with an Elaborate Buffer." U.S. Patent No. 6,148,534 (2000).

Love, A. E. H. *A Treatise on the Mathematical Theory of Elasticity.* 4th ed. New York: Dover, 1944.

Pellegrino, S., ed. *Deployable Structures.* Vienna: Springer, 2001.

Pellegrino, S., and S. D. Guest, eds. *IUTAM-IASS Symposium on Deployable Structures: Theory and Applications. Proceedings of the IUTAM Symposium Held in Cambridge, U.K., 6–9 September 1998.* Dordrecht: Kluwer, 2000.

Petroski, Henry. "Deployable Structures." *American Scientist,* March–April 2004, 122–26.

Ramirez, Anthony. "Turning Profits Hand over Wrist." *New York Times,* October 27, 1990.

Sturman, Catherine. "Dubai's Dynamic Tower Hotel: Top 7 Facts." *Construction,* May 16, 2020. www.constructionglobal.com/construction-projects/dubais -dynamic-tower-hotel-top-7-facts.

Timoshenko, Stephen P. *History of Strength of Materials: With a Brief Account of the History of Theory of Elasticity and Theory of Structures.* New York: Dover, 1983.

Vogel, Steven. *Cats' Paws and Catapults: Mechanical Worlds of Nature and People.* New York: W. W. Norton, 1998.

Volz, Frederick A. "Coilable Rule." U.S. Patent No. 2,156,907 (1939).

Wedesweiler, Madeleine. "World's First Rotating Skyscraper Planned for Dubai." *Commercial Real Estate,* April 16, 2019. www.commercialrealestate.com.au/news /worlds-first-rotating-skyscraper-planned-for-dubai-34567.

15. Anthropomorphic Models

Åkesson, Björn. *Understanding Bridge Collapses.* London: Taylor and Francis, 2008.

Baker, Benjamin. "Bridging the Firth of Forth." *Engineering,* July 29, 1887, 114, 116.

———. "Bridging the Firth of Forth." *Proceedings of the Royal Institution* 12, no. 81 (1887): 142–49.

———. *Long-Span Railway Bridges; Comprising Investigations of the Comparative, Theoretical and Practical Advantages of the Various Adopted or Proposed Type Systems of Construction, etc.* Philadelphia: Henry Carey Baird, 1868.

———. *Long Span Railway Bridges.* Rev. ed. London: E. and F. N. Spon, 1873.

Baker, William. "Burj Khalifa: A New Paradigm." Gordon H. Smith Lecture, Yale School of Architecture, January 26, 2012. http://video.yale.edu/video/burj -khalifa-new-paradigm.

———. "Princeton Engineering Lectures: Tall Buildings Lectures: Bill Baker." Video, February 4, 2014. www.youtube.com/watch?v=cSShh6bOFMk.

"Canadian." "The Quebec Bridge and the Forth Bridge." *Engineering News,* October 10, 1907, 391.

Frontinus, Sextus Julius. *The Two Books on the Water Supply of the City of Rome.* Translated by Clemens Herschel. Boston: New England Water Works Association, 1973.

Gray, Michael, and Angelo Maggi. *Forth Bridge: Evelyn George Carey.* Milan: Federico Motta, 2009.

Isaacson, Walter. *Leonardo da Vinci.* New York: Simon and Schuster, 2017.

Kuprenas, John. *101 Things I Learned in Engineering School.* New York: Grand Central, 2013.

Mackay, Sheila. *The Forth Bridge: A Picture History.* Edinburgh: HMSO, 1993.

McCullough, David. *The Great Bridge.* New York: Simon and Schuster, 1972.

Mueller, Benjamin, Marc Santora, and Cora Engelbrecht. "Discovering How Temperature and Touch Can Signal Nervous System." *New York Times,* October 5, 2021.

"A Novel Illustration of the Cantilever Principle." *Engineering News,* June 11, 1887, 385.

Paxton, Roland, ed. *100 Years of the Forth Bridge.* London: Thomas Telford, 1990.

Petroski, Henry. "An Anthropomorphic Model." *American Scientist,* March–April 2012, 103–7.

———. "Dorton Arena: On the Occasion of Its Fiftieth Anniversary." *American Scientist,* November–December 2002, 503–7.

———. *To Engineer Is Human: The Role of Failure in Successful Design.* New York: St. Martin's Press, 1985.

Phillips, Philip. *The Forth Railway Bridge; Being the Expanded Edition of* The Giant's Anatomy. Edinburgh: R. Grant and Son, 1890.

———. *Sketches of the Forth Bridge; or, The Giant's Anatomy, from Various Points of View.* Edinburgh: R. Grant and Son, 1888.

Vitruvius. *The Ten Books on Architecture.* Translated by Morris Hicky Morgan. New York: Dover, 1960.

Webster, Nancy, and David Shirley. *A History of Brooklyn Bridge Park: How a Community Reclaimed and Transformed New York City's Waterfront.* New York: Columbia University Press, 2016.

Westhofen, W. "The Forth Bridge." *Engineering,* February 28, 1890, 213–83.

Wills, Elspeth. *The Briggers: The Story of the Men Who Built the Forth Bridge.* Edinburgh: Birlinn, 2009.

Yeomans, David. *How Structures Work: Design and Behaviour from Bridges to Buildings.* Oxford: Wiley-Blackwell, 2009.

16. Visible and Invisible Hands

Agrawal, Roma. *Build: The Hidden Stories Behind Our Structures.* New York: Bloomsbury, 2018.

Anagnos, Thalia, Becky Carroll, Shannon Weiss, and David R. Heil. "Public Works for Public Learning: A Case Study." ASEE Annual Conference & Exposition, Atlanta, June 2013.

Baker, B. Letter to W. C. Unwin, dated December 1, 1887. Transcription by Roland Paxton.

Barnard, Jeff. "Lawn Chair Balloonists Recount Harrowing Flight over Oregon." *USA Today,* July 18, 2012.

Barry, Mark. "The Official Site of 'The Lawn Chair Pilot.'" www.markbarry.com /lawnchairman.html.

Bleys, Olivier. *The Ghost in the Eiffel Tower.* Translated by J. A. Underwood. London: Marion Boyars, 2004.

Brooke, David. "Book Review: One Hundred Years of the Forth Bridge." *Journal of Transport History,* September 1992, 200–201.

"Building Big: Educators' Guide." *PBS,* undated. www.pbs.org/wgbh/buildingbig /educator/index.html.

Chandler, Alfred D., Jr. *The Visible Hand: The Managerial Revolution in American Business.* Cambridge, Mass.: Harvard University Press, 1977.

Chang, Kenneth. "Balloon Ride to Offer Expansive View, for a Price." *New York Times,* October 23, 2013.

Consortium of Universities for Research in Earthquake Engineering. "December: The Golden Gate Bridge Outdoor Exhibition." In *2013 CUREE Calendar.* Richmond, Calif.: CUREE, 2013.

Gillispie, Charles Coulston. *The Montgolfier Brothers and the Invention of Aviation, 1783–1784.* Princeton, N.J.: Princeton University Press, 1983.

Golden Gate Bridge Highway and Transportation District. "Resisting the Twisting." Undated. www.goldengate.org/exhibits/resisting-the-twisting.

Guerrero, Susana. "Marin Wanted a BART Connection." *SFGATE,* July 16, 2021. www.sfgate.com/local/article/How-BART-almost-connected-to-Marin-by-way -of-the-16309661.php.

Klotz, Irene. "Ride with a View: U.S. Firm to Offer Balloon Excursions to Stratosphere." Reuters, October 22, 2013. www.reuters.com/assets/print?aid=USBRE 99L1BU20131022.

Kuhn, Thomas S. *The Structure of Scientific Revolutions.* 4th ed. Chicago: University of Chicago Press, 2012.

Levy, Matthys. *Why the Wind Blows: A History of Weather and Global Warming.* Hinesburg, Vt.: Upper Access, 2007.

Lewis, E. E. *How Safe Is Safe Enough? Technological Risks, Real and Perceived.* New York: Carrel Books, 2014.

Lewis, Peter R. *Beautiful Railway Bridge of the Silvery Tay: Reinvestigating the Tay Bridge Disaster of 1879.* Stroud, U.K.: Tempus, 2004.

Lund, Jay R., and Joseph P. Byrne. "Leonardo da Vinci's Tensile Strength Tests: Implications for the Discovery of Engineering Mechanics." *Civil Engineering and Environmental Systems* 18, no. 3 (2001): 243–50.

Macaulay, David. *Building Big.* New York: Houghton Mifflin, 2000.

Martin, T., and I. A. MacLeod. "The Tay Rail Bridge Disaster Revisited." *Proceedings of the Institution of Civil Engineers: Bridge Engineering* 157 (2004): 187–92.

Morgenstern, Joseph. "The Fifty-Nine-Story Crisis." *New Yorker*, May 25, 1995, 45–53.

Petroski, Henry. *To Engineer Is Human: The Role of Failure in Successful Design*. New York: St. Martin's Press, 1985.

Phillips, Philip. *The Forth Railway Bridge: Being the Expanded Edition of* The Giant's Anatomy. Edinburgh: R. Grant and Son, 1890.

Roberts, Siobhan. *Wind Wizard: Alan G. Davenport and the Art of Wind Engineering*. Princeton, N.J.: Princeton University Press, 2013.

Salvadori, Mario. *Why Buildings Stand Up: The Strength of Architecture*. New York: McGraw-Hill, 1980.

Smith, Adam. *An Inquiry into the Nature and Causes of the Wealth of Nations*. Dublin: Printed for Messrs. Whitestone, Chamberlaine, Watson, et al., 1776.

TA Corporation. "Golden Bridge." Undated. https://web.archive.org/web /20190409231746/http://talavn.com.vn/en/golden-bridge.

Vitruvius. *The Ten Books on Architecture*. Translated by Morris Hicky Morgan. New York: Dover, 1960.

Walker, E. G. *The Life and Work of William Cawthorne Unwin*. Sydney, Australia: G. Allen and Unwin, 1947.

Weidman, Patrick, and Iosif Pinelis. "Model Equations for the Eiffel Tower Profile: Historical Perspective and New Results." *Comptes Rendus Mécanique* 332 (2004): 571–84.

17. Overarching Problems

Kaplan-Leiserson, Eva. "Engineering Solutions." *PE Magazine*, March 2011, 20–23.

Macaulay, David. *Building Big*. New York: Houghton Mifflin, 2000.

Mainstone, Rowland J. *Developments in Structural Form*. Cambridge, Mass.: MIT Press, 1975.

Mannix, Nicholas C. "What Lies Beneath." *Structural Engineer*, August 2012, 44–46.

Mark, Robert. *Light, Wind, and Structure: The Mystery of the Master Builders*. Cambridge, Mass.: MIT Press, 1990.

National Academy of Engineering. *NAE Grand Challenges for Engineering*. Washington, D.C.: NAE, 2017.

Paterson, Mike, et al. "Maximum Overhang." Working paper, Dartmouth College. https://math.dartmouth.edu/~pw/papers/maxover.pdf.

Petroski, Henry. "Arches and Domes." *American Scientist*, March–April 2011, 111–15.

———. "Overarching Problems." *American Scientist*, November–December 2012, 458–62.

Rosenberger, Homer T. "Thomas Ustick Walter and the Completion of the United States Capitol." *Records of the Columbia Historical Society* 50 (1948–50): 273–322.

Taylor, Rabun. *Roman Builders: A Study in Architectural Process*. Cambridge: Cambridge University Press, 2003.

Tortorello, Michael. "Life with Pebbles and Bam Bam." *New York Times*, July 18, 2013.

18. Pyramids, Obelisks, and Asparagus

Acocella, Joan. "What the Stone Said." *New Yorker*, November 29, 2021, pp. 80–85.

Brien, James H. "Imhotep: The Real Father of Medicine? An Iconoclastic View." *Healio*, October 6, 2014. www.healio.com/news/pediatrics/20141203/imhotep -the-real-father-of-medicine-an-iconoclastic-view.

Brier, Bob. "How to Build a Pyramid." *Archaeology* 60, no. 3 (2007). www.archaeol ogy.org/0705/etc/pyramid.html.

Brier, Bob, and Jean-Pierre Houdin. *The Secret of the Great Pyramid: How One Man's Obsession Led to the Solution of Ancient Egypt's Greatest Mystery*. New York: Harper-Collins, 2008.

Curran, Brian A., Anthony Grafton, Pamela O. Long, and Benjamin Weiss. *Obelisk: A History*. Cambridge, Mass.: Burndy Library, 2009.

Dibner, Bern. *Moving the Obelisks*. Norwalk, Conn.: Burndy Library, 1991.

Edwards, James Frederick. "Building the Great Pyramid: Probable Construction Methods Employed at Giza." *Technology and Culture* 44, no. 2 (2003): 340–54.

Fontana, Domenico. *Della trasportatione dell'obelisco vaticano . . . .* Rome, 1590.

Galileo. *Dialogues Concerning Two New Sciences*. Translated by Henry Crew and Alfonso de Salvio. New York: Dover, [1954].

Gorringe, Henry H. *Egyptian Obelisks*. New York: Published by the author, 1882.

Hadingham, Evan. "A Nova Crew Strains, and Chants, to Solve the Obelisk Mystery." *Smithsonian*, January 1997, 22–32.

Houdin, Jean-Pierre, and Henri Houdin. *La construction de la pyramide de Khéops: Vers la fin des mystères?* Annales des ponts et chaussées 101. Paris: Ingénieur Science Société, 2002.

———. *La pyramide de Khéops*. Paris: Éditions du Linteau, 2003.

Isler, Martin. *Sticks, Stones, and Shadows: Building the Egyptian Pyramids*. Norman: University of Oklahoma Press, 2001.

Lehner, Mark. *The Complete Pyramids*. London: Thames and Hudson, 1997.

Lewis, M. J. T. "Roman Methods of Transporting and Erecting Obelisks." *Transactions of the Newcomen Society* 56 (1984–85): 87–110.

Petroski, Henry. "Moving Obelisks." *American Scientist*, November–December 2011, 448–52.

———. "Pyramids as Inclined Planes." *American Scientist*, May–June 2004, 218–22.

Risse, Guenter B. "Imhotep and Medicine—A Reevaluation." *Western Journal of Medicine* 144, no. 5 (1986): 622–24.

Stewart, Ian. "Pyramid Power, People Power." *Nature* 383 (1996): 218.

Verner, Miroslav. *The Pyramids: The Mystery, Culture, and Science of Egypt's Great Monuments*. Translated by Steven Rendall. New York: Grove Press, 2001.

Wier, Stuart Kirkland. "Insight from Geometry and Physics into the Construction of Egyptian Old Kingdom Pyramids." *Cambridge Archaeological Journal* 6, no. 1 (1996): 150–63.

Williams, John J. *The Williams's Hydraulic Theory to Cheops's Pyramid.* Albuquerque, N. Mex.: Consumertronics, 2005.

19. Moving with the Planet

Binczewski, George J. "The Point of a Monument: A History of the Aluminum Cap of the Washington Monument." *JOM* 47 (1995): 20–25.

Bing, Richard. "George Washington's Monument." *Constructor* 58 (1976): 18–25.

Burch, Gary A., and Steven M. Pennington, eds. *Civil Engineering Landmarks of the Nation's Capital.* Washington, D.C.: Committee on History and Heritage, National Capital Section, American Society of Civil Engineers, 1982.

Hoyt, William D., Jr. "Robert Mills and the Washington Monument in Baltimore." *Maryland Historical Magazine* 34 (1939): 144–60.

Lemieux, Daniel J., and Terrence F. Paret. "Monumental Challenge." *Civil Engineering,* December 2012, 50–67.

Lewis, Bob, and Vicki Smith. "East Coast Earthquake Closes Washington Monuments." Associated Press, August 24, 2011.

National Building Museum. "National Building Museum Is Open After a Post-Earthquake Inspection." Press release, August 25, 2011.

Petroski, Henry. "The Washington Monument." *American Scientist,* January–February 2012, 16–20.

———. "Why Buildings Remain Standing." *ChildArt Magazine,* April–June 2018, 8–9.

Pollard, Justin. *Buses, Bankers & the Beer of Revenge: An Eccentric Engineer Collection.* Stevenage, U.K.: Institution of Engineering and Technology, 2012.

Thompson, Ginger. "Quake Leaves Cracks in Washington Monument, Closing It for Now." *New York Times,* Aug. 25, 2011, p. A16.

Torres, Louis. *"To the Immortal Name and Memory of George Washington": The United States Army Corps of Engineers and the Construction of the Washington Monument.* Washington, D.C.: Office of the Chief of Engineers, 1984.

The Washington Monument. Washington, D.C.: Society of American Military Engineers, 1923.

20. Forces Felt and Heard

American Society of Civil Engineers, Metropolitan Section. "Statue of Liberty." Undated. www.ascemetsection.org/committees/history-and-heritage/landmarks/statue-of-liberty.

Cliver, E. Blaine. "The Statue of Liberty: Systems within a Structure of Metals." *Bulletin of the Association for Preservation Technology* 18 (1986): 12–23.

Cox, Trevor. *The Sound Book: The Science of the Sonic Wonders of the World.* New York: W. W. Norton, 2013.

Khan, Yasmin Sabina. *Enlightening the World: The Creation of the Statue of Liberty.* Ithaca, N.Y.: Cornell University Press, 2010.

Kipling, Rudyard. *The Day's Work.* Oxford: Oxford University Press, 1987.

McGeehan, Patrick. "Statue of Liberty Is to Reopen, Fittingly, by the Fourth of July." *New York Times,* March 20, 2013.

Petroski, Henry. *To Forgive Design: Understanding Failure.* Cambridge, Mass.: Harvard University Press, 2012.

Thomas, Dylan. *18 Poems.* London: Sunday Referee and Parton Bookshop, 1934.

Waldman, Jonathan. *Rust: The Longest War.* New York: Simon and Schuster, 2015.

Epilogue

Augustinians of North America. "St. Augustine and the Pear Tree: A Lasting Story." *Augustinian Vocations* (blog), January 19, 2017. https://beafriar.org/blog-archive/2017/1/19/st-augustine-and-the-pear-tree-a-lasting-story.

Fowler, Michael. "Discovering Gravity: Galileo, Newton, Kepler." Lecture notes, "Physics 152: Gravity," University of Virginia, 2002. https://galileo.phys.virginia.edu/classes/152.mf1i.spring02/DiscoveringGravity.htm.

Guicciardini, Niccolò. "Reconsidering the Hooke-Newton Debate on Gravitation: Recent Results." *Early Science and Medicine* 10, no. 4 (2005): 511–17.

Hall, A. Rupert. "'The Prime of My Age for Invention,' 1664–1667." In *Isaac Newton: Adventurer in Thought,* 30–64. Cambridge: Cambridge University Press, 1996.

Kaku, Michio. *The God Equation: The Quest for a Theory of Everything.* New York: Doubleday, 2021.

Keesing, Richard. "A Brief History of Isaac Newton's Apple Tree." Blog post, University of York, Department of Physics. www.york.ac.uk/physics/about/newtonsappletree.

Palter, Robert, ed. *The Annus Mirabilis of Sir Isaac Newton.* Cambridge, Mass.: MIT Press, 1971.

Index

Italicized page numbers refer to illustrations and tables.